The C.Y. O'Connor ERADE Village Foundation

Incorporating the Immunogenetics Research Foundation and the Institute of Molecular Genetics and Immunology Incorporated

E*ducation,* **R***esearch* **A***nd* **D***evelopment,* **E***mployment*

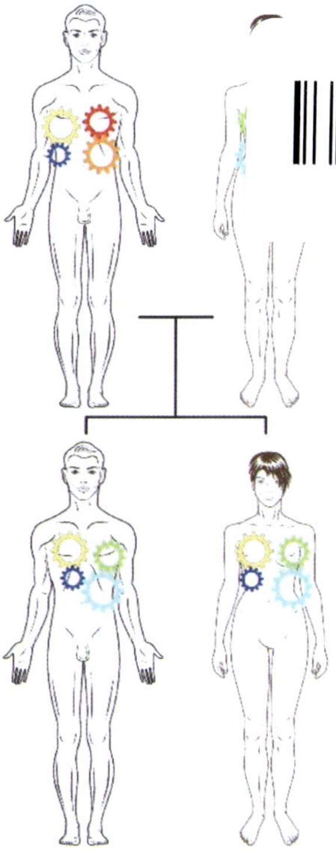

33rd Annual Report 2016

C Y O'Connor ERADE Village Foundation Inc.

Employment Research and Development
Employment Technology Incubator

Street Address		24 Genomics Rise Piara Waters WA 6112
Postal Address		PO Box 5100 Canning Vale South WA 6155
Telephone		(+61 8) 9397 1556

Key Contact		Professor Roger Dawkins
Position	Chairman	
Email		rldawkins@cyo.edu.au
Second Contact		Dr Sally Lloyd
Position	Director	
Email		slloyd@cyo.edu.au

Company Overview: CYO 'Connor ERADE Village Foundation seeks to harness the intellect of young Australian post graduates whilst maintain extensive collaboration with leading institutes overseas.

Recently expanding into multidisciplinary science in fields such as bioenergy, information technology, natural products, recycling and water, the core technology is interrogation of genetic information.

The aims include understanding of human evolution in the context of disease including retroviruses, identification of new genetic markers via genomic prospecting and development of diagnostic tests for inherited traits in numerous species of medical and commercial interest such as cattle, dogs, fish and invertebrates.

Product/Service Information: The CY O'Connor Education Research and Development Employment (ERADE Village) Foundation is a charitable Foundation that conducts independent and sponsored research and administers a federally funded technology incubator.

The Foundation has available low-cost two story R&D studios including accommodation if required, central laboratory facilities, conference rooms, visitor accommodation and sophisticated intranet and broadband connection.

Fellowships are awarded to outstanding researchers to assist them in commercialisation of their R&D.

Industry and government funding have established the Village as a multi-million dollar R&D and commercial centre following a model that is being replicated at the Albany Research Institute

Published by Nearurban LLC, 1603 LBJ Freeway,
Suite 280, Dallas, TX 75234
Tel (972) 974 7130, Fax (214) 393-4001
Email : admin@nearurban.net

Manufactured and printed by CreateSpace.

ISBN: 978-0-9864115-3-3

Title and Cover image: From R.L.Dawkins' "*Adapting Genetics – Quantal Evolution After Natural Selection- Surviving The Changes To Come*" 2015 Nearurban Publishing ISBN 978-0-9864115-1-9 : **Reactive meshing dependent on hormonal and other environmental influences.** Coloured cogs represent ancestral haplotypes. The size and shape of cogs represent polymorphism or variant forms of the blocks. The two offspring are genotypically identical but the interactions are different as a consequence of the hormonal environment.

NU
Publishing

FRIENDS OF CYO'CONNOR VILLAGE

The Foundation is a charitable organisation supporting education, research and development and employment. Without the help of friends of the Foundation it would not be possible to achieve all these objectives, including exciting new scientific discoveries. The Board thanks its many friends for their generous donations, bequests, collaborations, advice and assistance during 2015.

I/we (Name/s) .... (Tel)...............

Of (Address)... wish to:

Join the "Friends of CYO" 2015 ($55.00 pa) $................

Receive information about the Corporate Golf Day and other fundraising events ......

Be invited to Scientific Retreats ......

Receive information about arranging a bequest ......

Support the Foundation in other ways ......

Please accept my / our total payment of $................

Cheques/Money Orders payable to: The CYO'Connor ERADE Village Foundation.
OR
Please debit my **Credit Card** with the above total amount only. I understand that American Express, Diners, Mastercard or VISA are accepted. My card details are as follows:

Card Type: ....

Card No.: .... Expiry Date ...../.....

Cardholder's Name: .... Signature ....

POST TO: THE C. Y. O"CONNOR ERADE VILAGE FOUNDATION INC.
PO BOX 5100 CANNING VALE SOUTH WA 6155

AL & M DAWKINS FOUNDATION

Sponsors up to

$200,000 p.a. for Research & Development in WA

Residential scholarships and grants for:

FUTURE OF WA
BIOMEDICAL RESEARCH

in association with a university within WA if enrolled. Matching funding must be available or anticipated. Expressions of interest including track record should be addressed to:

CY O'Connor Village,
PO Box 5100, Canning Vale 6155

Closing Date

November 30th or May 31st each year

The AL & M Dawkins Foundation has been funded to support problem solving in Western Australia by:

(1) Defining important issues relating to urban development and planning and potential solutions thereto;
(2) Defining important issues relating to biomedical research and solutions thereto;
(3) Supporting prospective innovators in the above fields and by funding awards, scholarships, prizes and fellowships in association with CY O'Connor ERADE Village, or any branches thereof, in association with any university to which the CY O'Connor ERADE Village may be associated;
(4) Provide funding for patent costs to protect any intellectual property so developed.

Applications received by November 30 or May 31 each year will be evaluated by one of two expert panels which make recommendations to the Advisory Board,

The Board may seek further advice and will also consider:

(1) the relevance to multidisciplinary projects underway within the CYO
(2) the value of the individual to life within the village
(3) the capacity to leverage other funds
(4) university linkages
(5) potential IP

and will offer funding and accommodation for live-in researchers.

The town planning panel will be seeking to support innovative projects relating to water shortages and contamination, affordable housing, sustainable villages and isolated communities.

The biomedical panel will identify boutique projects which can be internationally competitive whilst relevant to local issues such as DNA matching, evolutionary genetics, livestock diseases and plant genomics.

Within 3 months, the administrators of the fund will advise applicants in the following categories

(1) successful
(2) for reconsideration when matching funding is confirmed
(3) invitation to reapply in the next cycle
(4) unsuccessful

Grants, generally of $50, 000 must include a stipend and rental but may also include funds for technical support, equipment, reagents, patents and publications. Travel grants will not be considered. The Foundation may decide to make no awards.

Funds will only be released after assignment of IP in return for a royalty negotiated on a case by case basis.

Table of Contents

Patron
Mr. S.L. Perron AM

Vice-Patrons
The Hon. Kim Beazley MA, MPhil
Senator Alan Eggleston BA, MRCS, LRCP, FRACGP
Mr. Max McDonald
Mr. Patrick Dawkins
Mr. Syd Martin

Board of Management

as of May 2015

Appointed Members
Professor Roger L. Dawkins - Research Director/Chairman
(MD,DSc,FRACP,FRCPA,FRCP)
Mr. Anthony Lloyd - Vice Chairman
(BSc (Ag)(Hons)MBA,LLB)
Dr Sally Lloyd - Deputy Research Director
(BSc (Hons) PhD)
M. Renny Lee-Steere - Deputy Vice Chairman
(EMBA)
Mr Edward Turner - Honorary Secretary
(FPIA, MTP, ATARP)
Ms Sue Lester - Deputy Honorary Secretary
(BSc)
Mr. Wayne Cox - Honorary Treasurer
(MTax, DipFS(FP), CPA)
Mr Jason Wallis -Deputy Honorary Treasurer
(BBus, R.E.Lic)

Elected Members
Mr Ben Dawkins (LLB)
Mr Roger Stubbs (AM BA, Grad Dip URP)
Mrs Sharon Baker
Mr Alan Peggs

Co-opted Members (by Invitation)
Dr Sam Dawkins, BSc, PhD
Mr Brian Walker

Life Members
Dr BJ Stewart, BVSc (Hons), Dip Ag
Mr John Heydon, BCom, CA, FTIA
Mr David GA Gibb, BA (Hons), AIBMS, AAIMS
Mr Lindsay Wolridge

Executive Reports

Chairman's Report Professor Roger L. Dawkins

Innovation and Akubra Hats

A visitor to Australia might conclude that there are two national priorities;

1. Innovate in electronics by having entrepreneurs wear jeans and T shirts,
2. Create Food Bowls by putting Akubra hats onto politicians and investors

and might wonder why we have accepted optimistic visions of "food bowls for the world" whilst, at the same time, permitting the withdrawal or diversion of funding for agricultural innovation and education. This paradox is all the more surprising given Australia's history of problem solving in agriculture. Patently, any realistic prospects for sustained export will be dependent on reinvigorating internationally competitive research and development. Innovators need Akubras!

Food will have to be safe and demonstrably so. Thus, success will require ongoing monitoring and clinical trials – ongoing because the inputs are ever changing as the environment varies and economies cycle. Consumers need convincing whether they worry about fat, sugar, GMO, insecticides, additives or "organic farming". Witness the contributions to the Global Food Forum summarized in The Australian in early 2016.

Obviously, food bowls require accurate identification and tracing of the food as well as credible assurances of safety and healthiness. It would be clever to discourage substitution and contamination whether at the paddock, the plate or the innumerable steps in between. Here in WA, we used to be very trusting of local food but following the application of CYOEVF's DNA ID patent to just some of these steps along the supply chain, we found that up to 50% of local produce had been substituted with inferior material of unknown provenance. Not only does this ruin reputations, encourage false claims and expose the public, but, worse for the future, it invalidates breeding programmes based on consumer demand. Basic research into healthy food becomes compromised and rather pointless in the absence of genetic tracing.

CYOEVF has to spend too much time and money ensuring provenance as a prerequisite for research. And, this is just in WA! What will happen to "food bowls" for regions which are truly expert at substitution and misrepresentation?
Anyone investing intellect or capital without research and development should hesitate. Genuine efforts may be thwarted as they have been previously.

Of course, excellent multidisciplinary and fundamental research is required to make Australia financially and sustainably productive whatever and wherever the food is produced. One of CYOEVF's very exciting and successful projects will improve the quality of northern beef. Others may increase investment returns by adapting financial and management schemes, but we will improve quality by identifying and patenting more of the relevant genes, spreading them by natural processes and monitoring provenance.

The CYOEVF is eclectic and is responding to local issues in the Peel region, including;

1. the need to retain nutrients on the farm,
2. milk production during summer,
3. improved solar irrigation,
4. greater than 20% unemployment of rural youth,
5. stealing of livestock and equipment,
6. unintentional protection of feral pigs and kangaroos rather than productive livestock,
7. mismanaged bush fire control,
8. lack of education for current and future farmers, etc.

All of these issues are worth addressing if innovation and food production are serious priorities.

The Foundation will provide matching funding up to $5m for the building and staffing of research and development facilities based on these principles.

The Australian of Thursday, November 12, 2015 says that venture capital investments in the USA are USD46,532 million, ie $50 billion, compared to $1,1652 in Israel but only $266 million in Australia. On a per capita basis the USA invests $147, Israel $145 and Australia only $11.

So the Foundation seeks support to increase innovation in food production!

Nearurban Planning

Our proposal for the Peel region includes some fairly simple suggestions to planners. Rather than increasing coastal sprawl, consider reinvigorating country towns and encourage Nearurban hamlets which could be attractive to innovators in Akubras. The idea is hardly radical. Witness the success elsewhere.

All The Best to Ted Steele

After almost 10 years here at CYOEVF, Ted is returning to the other side.

We are grateful for his many contributions to life in the village. These included lecture programmes, retreats, the books, scientific papers and ongoing analyses of the cattle records- not to forget driving us all around in the CYO bus.

I always enjoyed accompanying Ted on his pre-dawn rounds with the blue heelers. He will be sadly missed by friends in the tavern, the farms, the labs and the office.

We hope to see him back at future retreats to which he has a standing invitation.

I will be nominating Ted for life membership of the CYOEVF.

Research Director's Report Dr Sally S Lloyd

In 2015 CYO foundation research has made progress in measuring the healthiness of beef fat and in finding associated genetic markers. We continue to apply the successful haplotype strategies learned from the MHC to new fields and new technologies.

There has been growing public awareness this year of the positive role of low melting point, unsaturated fats in a healthy diet. Nina Tchekolz poular science book "A Big Fat Surprise" has shown how misguided dietary guidelines for reducing fat and replacing it with carbohydrates and sugar has been. Instead we should be reducing both saturated fats and simple carbohydrates and replacing them with mono unsaturated and polyunsaturated fats. In this environment, our current research focus on fat melting point in beef is topical and has been well received.

The fat melting point test developed at CYO has this year been used to establish differences in beef fat related to cattle breed, feed regime, specific cuts and individual pedigrees. In May, I was invited to the first international Wagyu conference held in Rockhampton, Queensland. The message that Wagyu fat was generally lower melting point, and therefore healthier, was well received. Our research showing that individual sires had a significant effect on the Tm of their progeny, and that high marbling genetics do not guarantee low Tm generated interest. Subsequently CYO was invited to present our research to the American Wagyu Association's. Professor Dawkins travelled to Florida in October to present the CYO testing methods for Tm and the associated haplotypes in Chromosome 19.

This research was submitted in two papers enclosed within:

Melting temperature of Wagyu beef fat significantly affected by sire selection.
Sally S Lloyd, Jose L Valenzuela, Dom Bayard, Scott de Bruin, Peter Gilmour, Edward J Steele and Roger L Dawkins. Submitted to Scientific Reports.

Haplotypes for Type, Degree and Rate of Marbling in Cattle are Syntenic with Human Muscular Dystrophy.
Sally S Lloyd, Edward J Steele, Jose L Valenzuela and Roger L Dawkins. Submitted to eLife.

Recent development of technology for next generation sequencing, and the amount of genetic information in publicly available databases has generated new opportunities to revisit the MHC haplotypes and use them to analyse the data now available. Prof Dawkins was invited to explain this approach as a chapter of a book on next generation sequencing, published as:
Sally S. Lloyd, Edward J. Steele and Roger L. Dawkins (2016). Analysis of Haplotype Sequences, Next Generation Sequencing - Advances, Applications and Challenges, Dr. Jerzy Kulski (Ed.), InTech, DOI: 10.5772/61794. Available from: http://www.intechopen.com/books/next-generation-sequencing-advances-applications-and-challenges/analysis-of-haplotype-sequences

I'd like to thank Dr Steele for his valuable contributions to all of these projects, and for his role organising several stimulating retreats this year.

Vice-Chairman's Report Mr. Anthony Lloyd

Anthony Lloyd is a retired solicitor who joined the Board in 2011. He has an interest in international intellectual property law, a subject relevant to the Foundation as many of its scientific discoveries can be commercialised through the patent system.

He was born in the Western Australian wheat belt and initially obtained a Bachelor of Agricultural Science degree at the University of Western Australia. Subsequently he obtained a Masters in Business Administration and a Bachelor of Laws.

Mr. Lloyd has a continuing interest in the Foundation's research relevant to agricultural industries and a particular interest in applying the Foundation's research expertise to minimise frost damage to the state's grain industry.

Kikuyu Pasture Improvement Project in the Peel Region of Western Australia

Background

The Foundation is fostering strong links with agricultural industries in the Peel region. Local Dairy and Beef producers have identified the need to improve their Kikuyu pastures by the identification of superior varieties for this region.

Dairy and Beef are significant elements of the agriculture industry in the Peel Region, particularly in the Serpentine-Jarrahdale, Murray and Waroona shires. Kikuyu Grass predominates as a summer pasture species on the coastal sand-plain.

Kikuyu has been the summer pasture plant of choice in the region due to its capacity to stabilise light and sandy soils under heavy grazing pressure. Its extensive root system accesses subsurface moisture to 1.5 – 2.0 metres so that in many situations green feed is available in summer months. In other situations, its persistence enables a Kikuyu pasture to take advantage of summer rainfall events. Kikuyu's deep-rootedness and persistence results in soil improvement through additional soil carbon, which in turn promotes the growth of other pasture species. A further benefit is that deep-rootedness and increased soil carbon reduce nutrient export into sensitive wetlands and waterways.

Maximizing pasture production and minimizing supplemental feeding is a key factor in maintaining an economically viable and environmentally sustainable livestock industry. While Kikuyu pastures are established on many properties and provide valuable summer feed, their agricultural value is limited by its issues of palatability and digestibility. Dairy cows especially will not consume it in sufficient quantities to maximise milk production without supplemental feeding.

Kikuyu grass was introduced into Australia from eastern Africa in the 1920s and planted extensively using vegetative propagation on the Peel coastal plain as well as the South Coast and Margaret River areas of Western Australia. In 1970, a new variety (Whittet) based on an introduction from South Africa was released and is the sole commercially available seed available for pasture production.

Very little research has occurred within Australia to identify new Kikuyu strains from within the existing population. However, in the last decade it has been established that the original pastures have adapted to local environments and a program of collection, assessment and breeding in the context of local agricultural systems is likely to result in superior varieties leading to improved Kikuyu pastures.

The limited collection and assessment programs currently undertaken in Australia are confined to the East Coast, largely funded by and directed to the requirements of the turf industry.

In the absence of NIRS calibration, laboratory assays of Kikuyu pasture quality are slow and expensive ($150-300) per assay and are therefore a significant disincentive to pasture management and further research. After calibration the equivalent cost could fall to less than $50 per assay.

Project Objectives

The project objectives are:

a) To identify strains of Pennisetum Clandestinum (Kikuyu Grass) which are superior to commercially available strains :

   i) for dairy and beef dairy production in the Peel region by reason of their superior:
      - Palatability
      - Digestibility
      - Contribution to the animal energy requirements
      - Ability to access sub-surface moisture so as to
      - minimise supplemental summer feeding of cattle for a more sustainable and viable agriculture in the region; and

   ii) for their ability to remove nutrients from sub-surface

water thereby minimising nutrient flows into sensitive wetlands and waterways.

b) To develop and calibrate a near infra-red spectroscopy (NIRS) laboratory method for measuring pasture quality parameters in Kikuyu pastures so as to provide cheaper and quicker laboratory assays.

c) To train locally based staff in the laboratory proce- dures including DNA extraction, and Polymerase Chain Reaction (PCR) and Bi-directional Amplification of Complementary Duplicons (BACD) methodology.

d) To provide a platform for further research into Kikuyu pasture improvement

Preliminary Work on the Project

A literature review of the state of research in Australia and elsewhere has been undertaken.

The Foundation's researchers have conducted a pilot program of DNA extraction from locally collected Kikuyu Grass samples and testing using PCR protocols and the Foundation's BACD technology. Those preliminary results indicate consistent genetic differences between the commercially available pasture variety (Whittet) and collections from farming and non-farming sites on the coastal plain originating from plantings from material introduced from southern Africa prior to the introduction and release of the Whittet variety which occurred in 1970. There are also substantial differences within the collections derived from those original plantings. The preliminary results indicate genetic diversity has developed from within the pre-1970 kikuyu population from a narrow genetic base. This diversity is likely to be an adaptive response to the environmental conditions on the coastal plain.

The project group has begun the process of identifying particular plants or stands of Kikuyu pasture which appear to have a
different growth pattern or growth habit than nearby stands or plants. The project group intends to widen this search to involve as many farmers as possible in the region by means of

a media campaign.

Contact has been made with relevant researchers and industry personnel to seek co-operation and identify current research and development activity.

A plant collection program has been initiated. It is intended to collect geographically diverse samples from within the Peel Region and the South West of Western Australia.

Treasurer's Report — Mr. Wayne Cox

2015 was fiscally a very good year for the Foundation. As a result of the restructure during 2014 our cash surplus jumped to $345,506 from a deficit of $51,791 the prior year. The sale of a parcel of GTG shares contributed $206,242 to the surplus, with the other noticeable change being the reduction in wages to $106,495 from $245,651.

This year also saw the landmark signing of a Joint Fellowship Agreement with Pardoo Beef Corporation Pty Ltd, which will contribute $200,000 over 4 years to the foundations research program.

Our cash reserves have grown to $1,174,406 due to the strong surplus, even after spending $180,000 improving the Incubators for sale and a further $62,000 protecting the Foundations intellectual property.

Our investment in GTG increased due to a sudden spike in the share price. Although we managed to sell down 3.0 m shares at 8c during the year, the end of year share price retreated to 2.5c, providing $325,417 in GTG shares at 31 December 2015

CYO Foundation Members Profiles & Reports

Dr. Ted Steele

Ted Steele is a cellular and molecular immunologist of some 40 years standing who rejoined the CYO Foundation in 2014 after a 3 year absence. His main work in recent years has been on the mechanism of somatic hypermutation of antibody variable region genes and on rapid evolutionary mechanisms. The main theme running through all this work has been the role of RNA intermediates and reverse transcription in the generation of genetic diversity. Recently this has led to the analysis of molecular mechanisms in the early stages of cancer development. He has a deep fascination with the whole topic of ancestral haplotypes and hopes to pursue his interests around this theme at CYO.

So we end another hectic year at CYO Foundation! There have been a number of achievements with resulting publications.

Origins of MHC Polymorphism: Haplotypic Structure of the Immunoglobulin (Antibody) Heavy Chain Variable Region Locus (IGHV)?

As previously reported this project attempts to unify several fundamental concepts in immunology. However as also reported the germ of the ideas morphed into a surprising discovery. A direct comparison of the extent of genetic diversity (polymorphism) between the ancestral haplotypes found at the MHC locus and attempts to measure the same type of megabase length polymorphism at the IGHV locus, concluded that IGHV may be many orders of magnitude *more polymorphic* than the MHC! – this is very significant as the MHC been considered hitherto the *most polymorphic region in the genome*.

The analysis of the data and critical arguments have now been published in the peer reviewed journal *BioEssays* (Steele and Lloyd 2015). It strongly implies that Soma-to-Germline feedback is likely to be *commonplace* for the genes encoding antibody variable regions. i.e. somatic antibody V mutants, stimulated and selected during the course of an immune response to infectious diseases are transmitted to progeny, genetically preparing them for protection if the disease persists in the family lineage. The main recommendation is that the 3-generation pedigree strategy, so successful in establishing segregation and stability of MHC ancestral haplotypes (and MRIP haplotypes on bovine chromosome 19) should now be applied to IGHV.

As we go to Press there has been a significant technical development which should accelerate the unravelling of the complexity of the IGHV locus – and provide direct proof for the Soma-to-Germline concept as it applies to the mammalian antibody genetic system.

A paper has just appeared in *Nature* by groups led by Dieter Egli of Columbia University and Nissim Benvenisty of the Hebrew University Jerusalem (Sagi et al 2016). They have found in certain lines of human embryonic stem cells (ESCs), that about 1 in 100 will remain haploid. Such haploid ESCs have been isolated and two cell lines have been established which carry only one copy of each human chromosome.

This development is important because the human IGHV 1 Mb locus (and others similar to it) has numerous highly similar V elements (both functional and non-functional) which are very difficult to actually sequence using current short-read methods (200 bp) typical of Next Generation Sequencing (NGS) technology. However, the ESCs being haploid means half of the battle of haplotype and sequence assembly has been beaten. If these ESC-IGHV regions can be sequenced and more easily assembled using very long read methods (10,000 bp) now coming on the market (Single Molecule Real Time, SMRT) by Pacific Bioscience then it is possible to imagine numerous IGHV haplotypes being monitored at megabase length and followed as they segregate through 3-generation families. If Soma-to-Germline feedback is as frequent as the current data suggest (Steele & Lloyd 2015) then the expected segregation will *very often not occur* as new somatic V mutants are inserted into individual germlines at each sperm-oocyte conception.

Cosmic Evolution and Panspermia

Unexpected developments can be the spice of life!

For many years I have followed the work of Fred Hoyle and Chandra Wickramasinghe who have marshalled the wide ranging and extensive evidence that life forms, microorganisms and viruses, from comets in our Solar System are daily infecting the Earth's biosystems. Over the long haul these cosmic genetic introgressions are driving biological evolution on Earth. There is much evidence they bring to bear – but the most prosaic is the obvious – collection systems from terrestrial balloon launches over the past 20 years show the presence of a wide variety of microorganisms and viruses on the sterile and protected collection pads. These viruses, bacteria and fungi have arrived from space and not by uplifting air currents from Earth. Similar unequivocal fossil-like forms of similar microorganisms have been shown earlier in meteorites. I recommend that sceptical colleagues both read *and then confront* the evidence for these conclusions for themselves. The best starting references are in a clear recent paper (Wickramasinghe et al 2015) and a wonderful extremely well written book by Chandra Wickramasinghe (Wickramasinghe 2015).

I had been developing certain ideas on Cosmic Evolution, Lamarck and Ancestral Haplotypes for some time. After recently re-confronting the Hoyle-Wickramasinghe (H-W) data again I wrote to Chandra Wickramasinghe, who not only warmed to the idea, but sent back a draft paper in reply! This paper is now published in the *Journal of Astrobiology and Outreach* (Wickramasinghe and Steele 2016).
In our view a cosmic origin for suddenly appearing virulent viral diseases like Zika should now be "the first cause of suspicion" as it is consistent with a wide body of specific data and evidence assembled by Wickramasinghe and Hoyle and their other colleagues over the past 40 years (Wickramasinghe 2015; Wickramasinghe et al 2015).

The current Zika virus has now been confirmed as the causative agent of microcephaly and infections in brain and placenta tissues of congenitally infected newborns in South America (Mlakar et al 2016; Martines et al 2016). That the virus has been isolated from semen in an infected male, as well as other media reports of sexual transmissions, may constitute *prima facie* evidence of Soma-to-Germ-line introgression of viral genes. So unlike HIV which is sexually transmitted and attacks the immune system, this non-retroviral single stranded RNA Flavivirus has at least one mosquito vector, infects humans and probably other higher non-human primates (and other animals), is sexually transmitted and attacks the central nervous system.

A dot-point summary of our main argument has several features:

- We believe the majority of scientists are both objective and willing to confront the wide range of data and evidence now cited.
- Since the current author has in fact done this for the data assembled by Wickramasinghe and Hoyle he urges others to follow suit.
- The assembled evidence is extensive, specific (e.g. stratosphere collections of cosmic micro-organisms and viral-like particles) and multifactorial (e.g. microorganisms embedded as fossils in meteorites, astrophysical analyses of comet emissions and chemistry etc,) such that any objective scientist confronted with the data would conclude an extra-terrestrial origin of all agents and phenomena identified.
- Since representatives of all microorganism groups have already infected the Earth on many past occasions (cosmic time scales of millions of years since four billions years ago) the prior existence on Earth of a particular virus (e.g. identified as "Zika") is not an argument against the most recent virulent form arriving from space.
- The current neo-Darwinian paradigm is both Earth-centred and obsolete. Evolutionary thinking, and thus biomedical science, needs a re-casted perspective to include all types of adaptations, cosmic biology sources, gene transfer modes and selective survival strategies operating since time immemorial dating back at least 13 billions years to the earliest vestiges of the visible Universe.

The Zika virus could have been invented by Hollywood central casting – but it is not a fantasy and the epidemic is gathering steam. We fear that continued adherence to manifestly wrong theories places mankind at great risk, unable to either understand or rationally respond to current and future biomedical challenges.

References

Martines RB, Bhatnagar J, Keating MK, et al. (2016) Notes from the Field: Evidence of Zika Virus Infection in Brain and Placental Tissues from Two Congenitally Infected Newborns and Two Fetal Losses -Brazil, 2015.MMWR Morbidity Mortality Weekly Report 2016;65 (Early Release) : 1–2. DOI: http://dx.doi.org/10.15585/mmwr.mm6506e1er).

Mlakar J, Korva M, Tul N, et al. (2016) Zika Virus Associated with Microcephaly New Engl J Med February 10, 2016, at NEJM.org. DOI: 10.1056/NEJMoa1600651.

Steele EJ, Lloyd SS. (2015) Soma-to-germline feedback is implied by the extreme polymorphism at IGHV relative to MHC. BioEssays. 37 : 557 – 569.

Steele EJ, Lindley RA, Blanden RV (1998) Lamarck's Signature: How Retrogenes are Changing Darwin's Natural Selection Paradigm. Allen and Unwin, Sydney

Sagi I, Chia G, Golen-Lev T, Peretz M, et al 2016 Derivation and differentiation of haploid human embryonic stem cells. *Nature* Published Online 16 march 2016 doi:10.1038/nature17408

Wickramasinghe C. (2015) The Search for Our Cosmic Ancestry. World Scientific, Singapore & New Jersey.

Wickramasinghe NC, Steele EJ (2016) Dangers of adhering to an obsolete paradigm: Could Zika virus lead to a reversal of human evolution? J. Astrobiol. Outreach.4:1 http://dx.doi.org/10.4172/2332- 2519.1000147

Wickramasinghe NC, Tokoro G, Wainwright M. (2015) The transition from Earth-Centred Biology to Cosmic Life. J Astrobiol. Outreach 3: 1

Dr. Samuel Dawkins

After some years in commercial R & D, Sam has spent over a decade in academia working in the field of quantum optics, with goals including the development of the next generation of atomic clocks for probing the remaining mysteries of physics such as dark matter and unifying theories. Currently, Sam works as part of a global effort to realise a functional quantum computer, which many believe will underpin the next generation of information technology.

Single Atom Heat Engine

Back in 1959, charismatic physicist Richard Feynman spoke of the possibility of having tiny nano-sized machines do the bidding of humans at a scales inaccessible in everyday life. He challenged physicists of the time to think big about thinking small and to make a start towards making machines at the ultimate limit of single atoms…

I recently had the pleasure, together with a team from the Universities of Mainz, Erlangen and Kassel, of achieving the first demonstration of an engine at the single atom level. We trapped a single calcium ion in special electrical fields, designed not only to hold the ion in place, but to allow it to convert thermal energy into a forced oscillation, much like you would see with a piston in your car engine. However, if you came to the lab, you would just see a vacuum chamber surrounded by lasers, optics and electronics. Being at the atomic level, its components are quite different to your car engine; the heat is produced with lasers and electrodes and the conversion of thermal energy to mechanical work is achieved through the ion's interaction with the electric fields rather than the interaction in a car between gas, piston and chamber.

The power produced by the single atom heat engine is also tiny: parts in 10^22 (1 with 22 zeros) watts. While James Watt (Watt's what watts are named after) would have been disappointed with that power output for his 1781 steam engine patent, we were very happy to be able to observe that amount of power. In fact, expressed in terms of power-to-mass ratio, it's at about the same level as a car engine.

What's the point? Well, this engine can not propel your car, and it can only function in the special circumstances we were careful to provide. But it does demonstrate that engines at that level are possible and it also provides the first step towards something that could serve a useful purpose. Imagine perhaps many machines hurtling around in a little factory, building something atom by atom. Or perhaps little machines running through the human body attacking cancer cells… Or a huge team of little machines all pushing and pulling in unison, making something happen at our macroscopic level…

For more details of the science, please go along and read our paper at Science: http://science.sciencemag.org/content/352/6283/325, or look at the pre-print version at: http://arxiv.org/abs/1510.03681

Mr. Roger Stubbs

Roger Stubbs is a graduate of UWA and Curtin University, Roger has a background in education, town planning and consultancy.

Roger is deeply involved in Local Government; he was Mayor of Armadale for eight years and a former President of the local Government Association, also an Executive member of the Australian Local Government Association. In 2002 Roger became a Member of the Order of Australia for service to local government through affiliated professional associations to urban planning, particularly through the development of environmental management policies and to the community of Armadale.

Planning for the new Peel CYO Foundation Laboratory

In 2015 The Peel Development Commission launched an ambitious strategy designed to promote the rural economy within its region. The strategy recognised the potential for intensive agricultural enterprises to exploit the region's proximity to large markets aided by an excellent transport system. Central to the strategy is the provision of an enhanced water supply and the development of a business park where secondary processors and distributors may be located. Of great interest to the Foundation was the recognition that high quality research is necessary if the strategy is to succeed.

Given that the recent focus of its research program has been upon the genetics of cattle, the Foundation immediately recognised an opportunity for it to play an important role within the Peel Region. Accordingly, the search began for an appropriate site to where the laboratory could be relocated. Following discussions with the Peel Development Commission and the Shire of Murray the Foundation identified the Recreation Reserve within the North Dandalup town site as its preferred location.

The Reserve was favoured because of its strategic location on the South West Highway and because of the level of services available in the town site. Further, its proximity and easy access to a number of both beef and dairy enterprises will facilitate the core research program. The Reserve also has an under utilised district hall which has the potential as a venue for seminars, retreats and conferences.

The Peel Development Commission's recognises that much of the proposed intensive agriculture will be horticulture based. If it is to succeed in an environmentally sensitive area, then effective management of water and nutrients will be critical. The Foundation recognises that its laboratory and expertise may play an important role, directly, and by attracting other professionals with particular expertise.

At the request of the Shire of Murray, the Foundation has prepared a master plan for the reserve. The master plan indicates the construction of three buildings. One building to house the laboratory, another to provide a number of serviced offices for those offering professional services to the rural community. A third building is proposed to function as a high tech men's shed where innovative people from the community may develop their projects. The master plan also provides for the renovation of the district hall and means for the reserve to better connect with what is an expanding community.

The master plan has received the support of the Shire of Murray but before it can enter into a lease agreement with the Foundation, it must meet its statutory obligation to advertise the master plan for public comment. Early indications are that

the master plan will receive the support of the community.
Once final approval is given, the Foundation must consider the best way of funding the construction of its new facilities. The Peel Development Commission has made a submission to the Royalties for Regions program for a series of grants, including a proposal for a research facility. The Foundation has met with a number of consultants who are undertaking due diligence of the proposal on behalf of the Department of Regional Development. Lottery west may provide a further opportunity for funding as the Foundation's proposal is generally consistent with its guidelines. Some capital will be available from the sale of the Foundation's incubator units in Canning Vale but the Foundation recognises the importance of preserving it as a means of meeting the ongoing costs of the research facility.

Mr Brian Walker — Fund Raising Report

Over the past 12 months there have been several events which proved to be successful in raing funds for the CY O'Connor Foundation, and no doubt these will be repeated into this year along with additional fund raising efforts.

Highlights over the year include:

The CY O'Connor Foundation charity golf day held at Gosnells golf course on 23rd September 2015.

The weather Gods certainly heeded the call with a perfect day for golf being provided which saw 8 teams take to the course. As I watched some of the shots I thought to myself more good weather was needed. So despite some ordinary shots being played (mine in particular) everyone had a wonderful day.

Our thanks go to Mr Tony Lloyd who MC'd the day but more importantly to our sponsors who made the day the success it was – with over $8,000 raised.

The overall winners were Team WFI (Wesfarmers Federation Insurance) comprising Colin Newby, Geoff Richards, John

Peddie and Brett Rumford† with a score of 49.75, followed by the Goldfather Corporation with 56, and Westpac Private Bank with 62.

Our Sponsors for the day were:

- Wallis property
- Westpac Private Bank
- Goldfather Corporation
- CY O'Connor Village Pub
- Lee-Steere & Co
- Oldfield Knott Architect

Always stand for your elders and follow the road rules

Hole Sponsors were:

- Knight Frank
- Sovereign Building

Donations:

- Joe Porter
- Primewest Management

Guys, did you see that – straight down the middle. Guys, GUYS!

A special thanks to James Wilkinson, General Manager, Gosnells Golf Course for his assistance in preparing the day and the donation of the Golf Buggies for use on the day.

† Brett Rumford is a close mate of Adam Scott, who has both offered him much advice etc. on his run to the 2103 Masters Green Jacket at Augusta National.

The CY O'Connor Foundation Charity Auction

October 2015 held at the CYO O'Connor Village Pub at the Dinner for the October Wagyu Retreat. Hats off to John Dawkins and the staff who provided a wonderful setting along with servicing their very own Paddock Plate Village Green Long Fed Wagyu & F1 Akaushi kebabs as part of the entree followed by Village Green Sirloin with Pumpkin ratatouille, green beans, crisp polenta with a balsamic jus.

Guests were entertained with being able to have caricatures of themselves drawn throughout the evening and these were in included in the main event being the Charity Auction.
The food and wine left everyone in good spirit for the Charity auction and with the auctioneer John Garland enticing the crowd several items saw bids exceeding expectations. Overall a very successful evening with over $8,000 raised.

Throughout the year the committee has raised over $18,000 for the Foundation and I would like to thank the committee members for their contribution and in particular the efforts of Dr Sally Lloyd, Dr Ted Steele, Roger Stubbs, John Garland, Tony Lloyd, Sharon Baker, and the unwavering support of Prof Roger Dawkins.

Mr Robert Giura

Dairy farmer, Keysbrook

Issues Affecting the Peel Region

1. Youth Unemployment, estimated to be as high as 20 per cent due to limited training facilities and job opportunities within the Peel Region.
2. Crime, particularly theft including from rural properties which can be difficult to secure adequately.

Issues Specific to Agriculture and Rural Landholders within the Peel Region

1. Increasing average age of farmers; younger generation choosing less demanding and more financially secure careers. Most farms in the Peel Region are family run operations and in many cases without succession plans for following generations to continue farming.
2. Feral animals; landholders adjoining Crown land and National Parks in particular are susceptible to ongoing damage to fencing, crops, pastures and spread of parasites from animals such as feral pigs, foxes and rabbits to a lesser extent are more widespread throughout the region. Native animals such as kangaroos can also cause considerable damage to fencing.
3. Unmaintained road verges; many years of accumulated vegetation on verges increases the risk

and intensity of fire damage to fences and infrastructure on rural properties and promote the spread of noxious and nuisance weeds to adjacent properties. Better maintenance and reduction of vegetation on roadside verges by councils could in some circumstances reduce the risk of fires crossing roads and assist with bringing fires under control.

4. Drainage; many low lying areas within the coastal plain of the Peel Region lack an integrated drainage system which results in flooding during periods of heavy rainfall leading to loss of productivity, poor drainage also causes salinity with longer term productivity losses in these areas.
5. Water sources; many farmers within the Peel Region rely wholly or partly on superficial aquifers to provide water for stock and domestic purposes, the increasingly drying climate is reducing the recharge of these aquifers and resulting in lowering water tables, placing greater pressure this resource.

Financial Statements

IMMUNOGENETICS RESEARCH FOUNDATION WA INC
T/A
THE C.Y. O'CONNOR ERADE VILLAGE FOUNDATION
67 161 830 965

FINANCIAL REPORT
FOR THE YEAR ENDED
31st December 2015

PASCOE PARTNERS
CERTIFIED PRACTISING ACCOUNTANTS
Liability limited by a scheme approved under
Professional Standards Legislation

IMMUNOGENETICS RESEARCH FOUNDATION WA INC
T/A
THE C.Y. O'CONNOR ERADE VILLAGE FOUNDATION
67 161 830 965

CONTENTS

Incubator Centre

IMMUNOGENETICS RESEARCH FOUNDATION WA INC
T/A
THE C.Y. O'CONNOR ERADE VILLAGE FOUNDATION
67 161 830 965
C Y O'CONNOR HI-TECH INCUBATOR CENTRE
DEPARTMENTAL TRADING, INCOME AND EXPENDITURE
FOR THE YEAR ENDED 31 DECEMBER 2015

	Note	2015 $	2014 $
C Y O'CONNOR HI-TECH INCUBATOR CENTRE			
OTHER INCOME			
Lease Rentals		198,707	199,614
EXPENSES			
Cleaning	2	-	440
Depreciation - Plant & Equipment	3	2,928	3,306
Depreciation - Buildings		26,233	23,565
Incubator Management Fees		22,575	141
Repairs & Maintenance	4	12,456	15,206
		64,192	42,658
SURPLUS (DEFICIT) FROM INCOME AND EXPENDITURE		134,515	156,956
Opening Balance 1st January 2015		1,200,654	1,043,698
Closing Balance 31st December 2015		1,335,169	1,200,654

Research and General

IMMUNOGENETICS RESEARCH FOUNDATION WA INC
T/A
THE C.Y. O'CONNOR ERADE VILLAGE FOUNDATION
67 161 830 965
RESEARCH & GENERAL FUND
DEPARTMENTAL TRADING, INCOME AND EXPENDITURE
FOR THE YEAR ENDED 31 DECEMBER 2015

	Note	2015 $	2014 $
RESEARCH & GENERAL			
Other Income		50,000	280
Donations/Bequests		179,943	131,411
Fund Raising		22,275	-
Meeting Registration Fees		9,140	7,130
Membership Fees		10,736	8,318
Laundry & Gaming Machine sales		1,610	2,593
		273,704	149,732
OTHER INCOME			
Interest Received		30,091	32,395
Reimbursements & Recoups		2,404	11,393
Profit on Sale of Non-current Assets		206,242	110
		238,737	43,898
EXPENSES			
Accountancy & Audit Fees		4,200	7,850
Advertising		2,252	80
Art Works Returned		-	299,658
Bank Charges		308	176
Cleaning		1,564	2,556
Consultancy Fees		2,750	7,819
Computer Expense		7,143	480
Depreciation - Plant & Equipment	3	18,587	23,266
Fees & Permits		324	1,594
Freight & Cartage		75	323
Fund Raising Costs		5,022	30
Innovation Award		30,000	30,000
Insurance		10,227	10,534
Legal Costs		5,823	518
Meeting/Visitors/Honoraria		16,254	12,743
Consumables & Lab Supplies		23,143	30,222
Motor Vehicle Expense		3,147	772
Patent Fees		62,181	-
Printing & Stationery		1,117	10,433
Repairs & Maintenance	4	-	5,284
Subscriptions		313	-
Superannuation Contributions		11,763	20,692
Telephone		6,415	11,453
Travelling Expenses		-	2,008

IMMUNOGENETICS RESEARCH FOUNDATION WA INC
T/A
THE C.Y. O'CONNOR ERADE VILLAGE FOUNDATION
67 161 830 965
RESEARCH & GENERAL FUND
DEPARTMENTAL TRADING, INCOME AND EXPENDITURE
FOR THE YEAR ENDED 31 DECEMBER 2015

	Note	2015 $	2014 $
Unrealised loss/ (gain) on Shares		(195,250)	640,667
Wages		106,495	245,651
Waste Disposal		934	1,159
		124,787	1,365,968
SURPLUS (DEFICIT) FROM INCOME AND EXPENDITURE		387,654	(1,172,338)
Opening Balance 1st January 2015		954,311	2,126,649
Transfer from Emergency Fund		5,000	-
Transfer from 7th MHC Evolution Meeting Fund		22,779	-
		27,779	-
Closing Balance 31st December 2015		1,369,744	954,311

Balance Sheet

IMMUNOGENETICS RESEARCH FOUNDATION WA INC
T/A
THE C.Y. O'CONNOR ERADE VILLAGE FOUNDATION
67 161 830 965
BALANCE SHEET
AS AT 31 DECEMBER 2015

	Note	2015 $	2014 $
TRUST FUNDS			
RESERVES	5	2,704,913	2,182,744
TOTAL TRUST FUNDS		2,704,913	2,182,744
Represented by:			
CURRENT ASSETS			
Cash on Hand		1	1
Petty Cash		82	183
Deposits on Building Contracts		-	72,439
Westpac Max 112867		16,068	39,510
Westpac Research Account 10-4752		3,174	1,872
Westpac Community Solutions Account 1645		23,140	14,175
Term Deposits	8	1,131,024	850,000
Accounts Receivable		22,316	26,951
Loans - Unsecured	6	10,000	-
GST Refundable		11,165	-
TOTAL CURRENT ASSETS		1,216,970	1,005,131
NON CURRENT ASSETS			
Fixed Assets			
Computer Equipment		364,971	364,971
Less: Accumulated Depreciation		(363,117)	(361,946)
		1,854	3,025
Motor Vehicles		13,500	13,500
Less: Accumulated Depreciation		(12,572)	(12,303)
		928	1,197
Office Furniture & Equipment		11,612	11,612
Less: Accumulated Depreciation		(10,942)	(10,774)
		670	838
Laboratory Equipment		507,296	507,296
Less: Accumulated Depreciation		(423,824)	(406,845)
		83,472	100,451
Incubators		55,124	51,392
Less: Accumulated Depreciation		(41,092)	(38,164)
		14,032	13,228
Incubator Buildings		1,287,296	1,105,241
Less: Accumulated Depreciation		(260,764)	(234,531)
		1,026,532	870,710
Total Fixed Assets		1,127,488	989,449
Investments			
Listed Shares in Public Companies	7	325,417	160,167
Patents		54,750	54,750
Total Investments		380,167	214,917

IMMUNOGENETICS RESEARCH FOUNDATION WA INC
T/A
THE C.Y. O'CONNOR ERADE VILLAGE FOUNDATION
67 161 830 965
BALANCE SHEET
AS AT 31 DECEMBER 2015

	Note	2015 $	2014 $
TOTAL NON CURRENT ASSETS		1,507,655	1,204,366
TOTAL ASSETS		2,724,625	2,209,497
CURRENT LIABILITIES			
Trade Creditors		1,575	9,034
Incubator Bond Monies	9	11,490	11,810
Amounts Withheld	10	6,647	4,649
GST Payable		-	1,260
TOTAL CURRENT LIABILITIES		19,712	26,753
TOTAL LIABILITIES		19,712	26,753
NET ASSETS		2,704,913	2,182,744

Notes

IMMUNOGENETICS RESEARCH FOUNDATION WA INC
T/A
THE C.Y. O'CONNOR ERADE VILLAGE FOUNDATION
67 161 830 965
NOTES TO THE FINANCIAL STATEMENTS
FOR THE YEAR ENDED 31 DECEMBER 2015

	2015	2014
	$	$

1 **Statement of Significant Accounting Policies**

The director of the trustee company has prepared the financial statements of the trust on the basis that the trust is a non-reporting entity because there are no users dependent on general purpose financial statements. The financial statements are therefore special purpose financial statements that have been prepared in order to meet the needs of beneficiaries.

The financial statements have been prepared in accordance with the significant accounting policies disclosed below which the director has determined are appropriate to meet the needs of members. Such accounting policies are consistent with the previous period unless stated otherwise.

The financial statements have been prepared on an accruals basis and are based on historical costs unless otherwise stated in the notes. The accounting policies that have been adopted in the preparation of these statements are as follows:

Property, Plant and Equipment

Property, plant and equipment are carried at cost, independent or director's valuation. All assets, excluding freehold land and buildings are depreciated over the useful lives of the assets to the trust.

Increases in the carrying amount arising on revaluation of land and buildings are credited to a revaluation reserve in shareholders' equity. Decreases that offset previous increases of the same asset are charged against fair value reserves directly in equity; all other decreases are charged to the income statement. Each year the difference between depreciation based on the revalued carrying amount of the asset charged to the income statement and depreciation based on the asset's original cost is transferred from the revaluation reserve to retained earnings.

The carrying amount of plant and equipment is reviewed annually by directors to ensure it is not in excess of the recoverable amount from these assets. The recoverable amount is assessed on the basis of the expected net cash flows that will be received from the asset's employment and subsequent disposal. The expected net cash flows have not been discounted in determining recoverable amounts.

Property, Plant and Equipment (PPE)

Leasehold improvements and office equipment are carried at cost less, where applicable, any accumulated depreciation.

The depreciable amount of all PPE is depreciated over the useful lives of the assets to the trust commencing from the time the asset is held ready for use.

Leasehold improvements are amortised over the shorter of either the unexpired period of the lease or the estimated useful lives of the improvements.

The accompanying notes form part of these financial statements.
This report is to be read in conjunction with the attached compilation report.

IMMUNOGENETICS RESEARCH FOUNDATION WA INC
T/A
THE C.Y. O'CONNOR ERADE VILLAGE FOUNDATION
67 161 830 965
NOTES TO THE FINANCIAL STATEMENTS
FOR THE YEAR ENDED 31 DECEMBER 2015

		2015 $	2014 $
2	**Cleaning**		
	Hi-Tech Incubator Centre	-	440
	Research & General Fund	1,564	2,556
		1,564	2,996
3	**Depreciation - Plant & Equipment**		
	Hi-Tech Incubator Centre	2,928	3,306
	Research & General Fund	18,587	23,266
		21,515	26,572
4	**Repairs & Maintenance**		
	Hi-Tech Incubator Centre	12,456	15,206
	Research & General Fund	-	5,284
		12,456	20,490
5	**RESERVES**		
	Emergency Fund	-	5,000
	7th MHC Evolution Meeting Fund	-	22,779
	C Y O'Connor Hi-Tech Incubator Centre	1,335,169	1,200,655
	Research & General Fund	1,369,744	954,310
		2,704,913	2,182,744
6	**Loans - Unsecured**		
	E J Steele	10,000	-
		10,000	-
7	**Listed Shares in Public Companies**		
	Shares in Listed Companies	9,762,500	12,012,500
	Less Provision for Diminution in Value	(9,437,083)	(11,852,333)
		325,417	160,167

Listed Shares in Public Companies include 13,016,667 shares held in Genetic Technologies Ltd. In 2004 the Foundation sold patents to Genetic Technologies Ltd. The agreement provided for the sale of the patents by the Foundation to Genetic Technologies Ltd for an amount of $15,000,000 and the granting of a licence by Genetic Technologies Ltd the foundation for the use of patents, at a cost of $2,500,000. The sale agreement provided the Foundation would receive net consideration of $12,500,000 to be Satisfied by 16,666,667 shares at an agreed price of 75 cents each. The shares were issued to the The Foundation on the 16th of June 2004. The shares were subject to escrow. Twenty percent of the total shareholding came out of escrow on each anniversary of the acquisition date. The ASX share price of Genetic Technologies Ltd at 31 December 2015 was 2.5 cents per share. Although the shareholding is considered a long term investment by the foundation, in accordance with normal accounting practice, the Foundation has provided $9,437,083 for the diminution in the value of the shares between the issue price and the market price at 31 December 2015.

The accompanying notes form part of these financial statements.
This report is to be read in conjunction with the attached compilation report.

IMMUNOGENETICS RESEARCH FOUNDATION WA INC
T/A
THE C.Y. O'CONNOR ERADE VILLAGE FOUNDATION
67 161 830 965
NOTES TO THE FINANCIAL STATEMENTS
FOR THE YEAR ENDED 31 DECEMBER 2015

		2015 $	2014 $
8	**Term Deposits**		
	Westpac Term Deposit 17-7599	125,000	125,000
	Westpac Term Deposit 18-6946	125,000	125,000
	Westpac Term Deposit 21-4134	125,000	250,000
	Westpac Term Deposit 21-4126	100,000	125,000
	Westpac Term Deposit 21-5671	125,000	125,000
	Westpac Term Deposit 21-9453	104,182	100,000
	Westpac Term Deposit 23-4151	100,000	-
	Westpac Term Deposit 23-4178	100,000	-
	Westpac Term Deposit 23-4186	100,000	-
	Westpac Term Deposit 21-1689	126,842	-
		1,131,024	850,000
9	**Incubator Bond Monies**		
	Auslanka	1,120	1,120
	LimeLight	720	720
	Better Care Medical Centre	1,360	1,360
	Youme Clothing	450	450
	Spicy Wok	1,200	1,200
	Make Your Mark Art Cafe	1,040	1,040
	Infinity Wealth Solutions	-	320
	Renov8withus Pty Ltd	1,400	1,400
	Indebiti Solutions Pty Ltd	1,400	1,400
	Alight Scent	1,400	1,400
	Mr Stubby	1,400	1,400
		11,490	11,810
10	**Amounts Withheld**		
	Amounts Withheld from Salary, Wages and Other Payments	2,406	1,284
	Employee Superannuation Payable	4,241	3,365
		6,647	4,649

Compilation Report

COMPILATION REPORT
IMMUNOGENETICS RESEARCH FOUNDATION WA INC
T/A
THE C.Y. O'CONNOR ERADE VILLAGE FOUNDATION
67 161 830 965

We have compiled the accompanying special purpose financial statements of THE C.Y. O'CONNOR ERADE VILLAGE FOUNDATION which comprise the balance sheet as at 31 December 2015, profit and loss statement and cash flow statement for the year then ended, a summary of significant accounting policies and other explanatory notes.

The specific purpose for which these special purpose financial statements have been prepared is to provide information relating to the performance and financial position of the incorporated body that satisfies the information needs of the director.

The responsibility of the Director

The director is solely responsible for the information contained in the special purpose financial statements and has determined that the basis of accounting used is appropriate to meet their needs and for the purpose for which the financial statements were prepared.

Our responsibility

On the basis of information provided by the director, we have compiled the accompanying special purpose financial statements in accordance with the financial reporting framework described in Note 1 to the financial statements and APES 315: Compilation of Financial Information.

Our procedures use accounting expertise to collect, classify and summarise the financial information, which the director provided, into a financial report. Our procedures do not include verification or validation procedures. No audit or review has been performed and accordingly no assurance is expressed.

The special purpose financial statements were prepared exclusively for the director. We do not accept responsibility to any other person for the content of the special purpose financial statements.

Name of Firm: Pascoe Partners
Certified Practising Accountants

Name of Partner: ______________________________
Wayne Michael Cox
MTax, Dip FS(FP), CPA, CTA

Address: Level 2, 100 Havelock Street West Perth

Dated:

The accompanying notes form part of these financial statements.
This report is to be read in conjunction with the attached compilation report.

Auditor's Report

The C.Y. O'Connor Erade Village Foundation
Independent Auditor's Report
to the Members of
The C.Y. O'Connor Erade Village Foundation
Incorporating
The Immunogenetics Research Foundation
and
The Institute of Molecular Genetics and Immunology Incorporated

---

We have audited the accompanying financial report, being a special purpose financial report, of The C.Y. O'Connor Erade Village Foundation (the association), which comprises the committee's report, the assets and liabilities statement as at 31 December 2015, the income and expenditure statement for the year then ended, notes comprising a summary of significant accounting policies and other explanatory information, and the statement by members of the committee.

Committee's Responsibility for the Financial Report

The committee of The C.Y. O'Connor Erade Village Foundation is responsible for the preparation of the financial report, and has determined that the basis of preparation described in Note 1 is appropriate to meet the requirements of the Associations Incorporation Act (1987) and is appropriate to meet the needs of the members. The committee's responsibility also includes such internal control as the committee determines is necessary to enable the preparation of a financial report that is free from material misstatement, whether due to fraud or error.

Auditor's Responsibility

Our responsibility is to express an opinion on the financial report based on our audit. We have conducted our audit in accordance with Australian Auditing Standards. Those standards require that we comply with relevant ethical requirements relating to audit engagements and plan and perform the audit to obtain reasonable assurance whether the financial report is free from material misstatement.

An audit involves performing procedures to obtain audit evidence about the amounts and disclosures in the financial report. The procedures selected depend on the auditor's judgment, including the assessment of the risks of material misstatement of the financial report, whether due to fraud or error. In making those risk assessments, the auditor considers internal control relevant to the association's preparation of the financial report that gives a true and fair view, in order to design audit procedures that are appropriate in the circumstances, but not for the purpose of expressing an opinion on the effectiveness of the association's internal control. An audit also includes evaluating the appropriateness of accounting policies used and the reasonableness of accounting estimates made by the committee, as well as evaluating the overall presentation of the financial report.

We believe that the audit evidence we have obtained is sufficient and appropriate to provide a basis for our audit opinion.

Qualification

In common with most charitable institutions receiving cash donations, the entity has determined that it is impractical to establish controls over cash donations before they are banked and recorded in the accounting system. As a result our audit was restricted to the amounts recorded in the accounting system.

Opinion

In our opinion, except for the effects on the financial report, if any that the above qualification may have on the financial report of The C.Y. O'Connor Erade Village Foundation, the report presents fairly, in all material respects, the financial position of The C.Y. O'Connor Erade Village Foundation as at 31 December 2015 and its financial performance for the year then ended in accordance with the accounting policies described in Note 1 to the financial statements, and the Associations Incorporation Act (1987).

The C.Y. O'Connor Erade Village Foundation
Independent Auditor's Report
to the Members of
The C.Y. O'Connor Erade Village Foundation
Incorporating
The Immunogenetics Research Foundation
and
The Institute of Molecular Genetics and Immunology Incorporated

---

Basis of Accounting and Restriction on Distribution
Without modifying our opinion, we draw attention to Note 1 to the financial report, which describes the basis of accounting. The financial report has been prepared to assist The C.Y. O'Connor Erade Village Foundation to meet the requirements of the Associations Incorporation Act (1987). As a result, the financial report may not be suitable for another purpose.

S. Williams

Steve Williams FIPA
Auditor

Dated this 8TH day of APRIL 2016

Auditors Independence Declaration
Under Section 307C of the Corporations Act 2001

To Members of The C.Y. O'Connor Erade Village Foundation

I declare that, to the best of my knowledge and belief, in relation to the audit of The C.Y. O'Connor Erade Village Foundation for the year ended 31 December 2015 there have been:

a) no contraventions of the auditor independence requirements of the Corporations Act 2001 in relation to the audit; and

b) no contraventions of any applicable code of professional conduct in relation to the audit.

S. Williams

Steve Williams FIPA
Auditor

Dated this 8TH day of APRIL 2016

Recent Foundation Publications

Ancestral Haplotypes

Front Cover

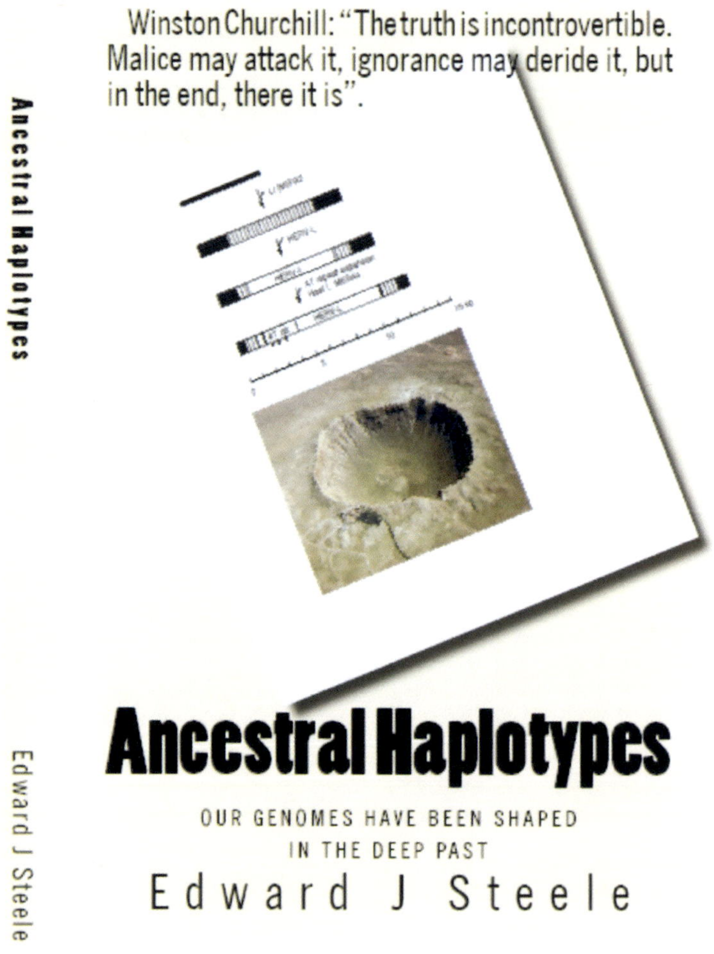

Back Cover

This book recounts a great scientific discovery. It tells the story of the leviathan effort to demonstrate that the structure of the DNA sequences in our chromosomes is not what it might seem. A naive view of evolution has invaded mainstream scientific thinking: that all our important genetic inheritance can be viewed under a simple Darwinian view of evolution based on recent and ongoing random gene mutation followed by 'natural selection'. To be more explicit, the neo-Darwinian paradigm based on Population Genetics and statistics. The discovery of "ancestral haplotypes" radically changes this simple genetic view. In most regions of the genome many of our important genes are very old, dating to mankind's migration out of Africa ca. 100,000 years ago. They occur in large segmentally duplicated clusters - chromosomal blocks which carry tens to hundreds of genes "frozen in time". These conserved blocks are shuffled at each generation to produce individual human beings, who, for the most part, are healthy and robust and well adapted. These are our "ancestral haplotype" sequences and they are the dominant feature of genomic structure, not the simple "beads on a string" model of traditional Population Genetics and neo-Darwinism. This book tells the scientific story of their discovery and the new picture of the genome it reveals.

NU Publishing

Barcode Area

We will add the barcode for you.

Made with Cover Creator

Adapting Genetics

FrontCover

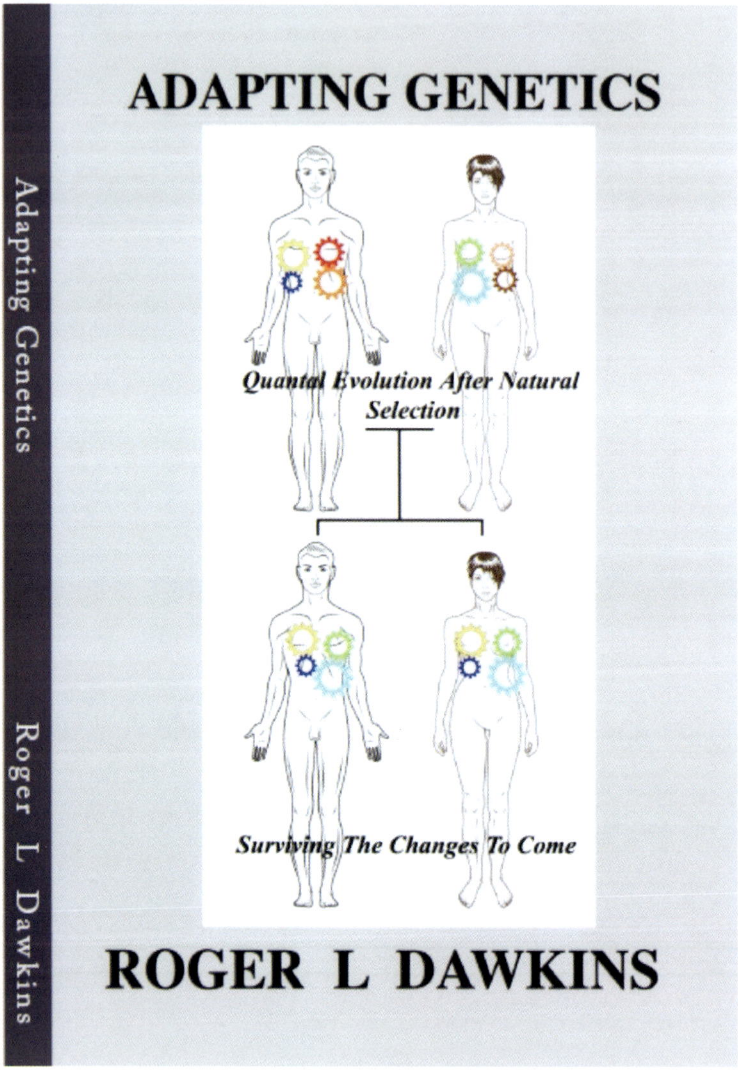

Back Cover

Genetics and evolution, as taught for a century, imply that survival is a lottery dependent upon random errors in copying the genome.

Quantal evolution refers to the partitioning of the genome to protect critical regulatory DNA sequences, including during the successive ***retroviral*** epidemics which defined human ***ancestors***. Trial and error selected pre-tested combinations which are faithfully inherited thereby allowing the new species to adapt by changing gears in the ***absence*** of natural selection or mutations. Heterozygous ***polymorphisms*** interact polymorphically, thereby creating additional options. There are clear implications for ***forensics***, ***livestock selection*** and ***transplantation*** inter alia.

Anamnestic evolution ensures individuality, diversity, protection from repeat epidemics and survival in the future.

" I enjoyed the book very much. It is a refreshing look at genetic evolution, which is really needed.

.....Professor Stephen Oppenheimer, author of "Out of Eden" and other books

" I could not stop reading it. How good it is to see that Professor Dawkins has not changed his enjoyable way of arguing."

....Professor Jon van Rood, leading HLA pioneer who's work helped establish the clinical and scientific study of the Major Histocompatibility Complex.

NU Publishing

Barcode Area

We will add the barcode for you.

Made with Cover Creator

A novel method of measuring the melting point of animal fats

A novel method of measuring the melting point of animal fats[1]

S. S. Lloyd,[2] S. T. Dawkins, and R. L. Dawkins

CY O'Connor ERADE Village Foundation, 11 Erade Drive, Piara Waters, Western Australia 6112

ABSTRACT: The melting point (T_M) of fat is relevant to health, but available methods of determining T_M are cumbersome. One of the standard methods of measuring T_M for animal and vegetable fats is the slip point, also known as the open capillary method. This method is imprecise and not amenable to automation or mass testing. We have developed a technique for measuring T_M of animal fat using the Rotor-Gene Q (Qiagen, Hilden, Germany). The assay has an intra-assay SD of 0.08°C. A single operator can extract and assay up to 250 samples of animal fat in 24 h, including the time to extract the fat from the adipose tissue. This technique will improve the quality of research into genetic and environmental contributions to fat composition of meat.

Key words: animal fat, beef, fatty acid, melting point

J. Anim. Sci. 2014.92:4775–4778
doi:10.2527/jas2014-7981

INTRODUCTION

The type of fat in the diet is as important as the quantity for maintaining good health. Diets that replace SFA with MUFA or PUFA are similar to or better than diets that replace SFA with carbohydrates. These effects can be seen in fasting glucose (Paniagua et al., 2007) and cholesterol profile (Mensink et al., 2003) and on coronary heart disease (Micha and Mozaffarian, 2010). The effect of MUFA and PUFA content on palatability and health is associated with their lower melting points (**T_M**). Even within the SFA, lauric acid has a lower T_M than stearic or palmitic acid (Knothe and Dunn, 2009) and a better effect on the ratio of total cholesterol to high-density lipoprotein cholesterol (Mensink et al., 2003).

Blood lipids improve when beef with high SFA content is replaced in the diet by beef with high MUFA content (Adams et al., 2010; Gilmore et al., 2011). Increased MUFA content also improves the flavor and texture of beef (Garmyn et al., 2011). Research into reducing the T_M of beef fat and increasing its MUFA and PUFA content is occurring (Pitchford et al., 2002; De Smet et al., 2004; Daley et al., 2010; Uemoto et al., 2010) but would benefit from a better method of measuring T_M.

Currently there are several fat T_M methods, of which the most commonly used is the slip point or open capillary method (American Oil Chemists' Society official method Cc 3b-92; AOCS, 1998). Other methods are less commonly used because they are more cumbersome, expensive, or inaccurate. All these methods lack the capacity for real-time multiple sample testing.

Therefore, an accurate T_M assay suitable for testing fat from large numbers of animals is desirable. Here we describe a method, using the Rotor-Gene Q real-time PCR Cycler (Qiagen, Hilden, Germany), which allows T_M measurements on up to 71 fat samples per 20-min run.

MATERIALS AND METHODS

Rotor-Gene Q Assay

The Rotor-Gene Q features rapid and precise temperature control, centrifugation, and real-time fluorescence detection that, although developed for high-resolution melting analysis of DNA, are exactly the components required to perform a unique measurement of fat T_M.

Twenty microliters of melted fat samples were transferred into 0.1-mL tubes. The tubes were then centrifuged at 9.8 × g and at 75°C for 2 min to collect the fat at the base of the tubes. The fat was solidified at 4°C for at least 12 h. Then, 10 μL of 26 μ*M* fluorescein dye in deionized distilled water was added to form a layer above the fat. The samples were conditioned at 25°C for 90 s in the Rotor-Gene Q assembly. During

[1]The research was supported by the CY O'Connor ERADE Village Foundation. The authors would like to thank Scott de Bruin, Dom Bayard, and J.R. Dawkins for fat samples from their cattle.
[2]Corresponding author: slloyd@cyo.edu.au
Received April 29, 2014.
Accepted July 22, 2014.

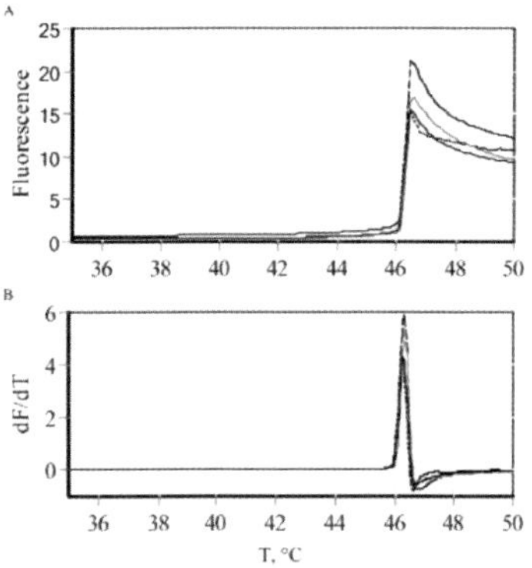

Figure 1. Fluorescence–temperature curves for 4 aliquots of rendered animal fat. Fluorescence is measured in relative units as a percentage of the maximum (saturated) signal. The derivative of fluorescence with respect to temperature (dF/dT) was calculated from the smoothed data using the Rotor Gene Q software.

the conditioning step, the photomultiplier gain was optimized using a tube loaded with dye only, which provided a reference fluorescence level.

Fluorescence detection in the Rotor-Gene Q is focused on the base of each tube. At the start of a run, the dye was above the fat and the fluorescence detection was low (see Fig. 1a for temperatures below 45°C). During the run, the temperature increased in 0.1°C steps with a rest time of 2 s at each step. At the T_M transition there was a sharp increase in fluorescence (Fig. 1a; at about 46°C). Gravitational and centrifugal forces pulled the dye down, through the liquefied fat, into the field of view of the detector, increasing the fluorescence signal. The inversion occurred because fat is less dense than the aqueous dye.

For convenience, and to ease automation, we define the T_M as the temperature at which the fluorescence signal changes most rapidly. This temperature, at the peak in the derivative of fluorescence with respect to temperature (shown in Fig 1b) is calculated automatically by the Rotor-Gene Q software.

Samples

We chose 3 standards within the T_M range of interest:

1. A sample of adipose tissue taken from the skirt fat of a Simmental steer slaughtered at 12 mo. We heated the adipose tissue to 90°C for 12 h and separated the melted fat from the water and solids by pipette.
2. Supafry (Goodman Fielder, NSW, Australia) is a commercially available rendered animal fat with a saturated fat content of 48% (wt/wt).
3. Allowrie Ghee (Fonterra, VIC, Australia) is clarified butter and has a saturated fat content of 66% (wt/wt).

Aliquots of each standard were stored at 4°C.

Fat samples were taken from 67 full-blood Wagyu steer carcasses slaughtered at an average age of 26.3 mo (SD = 1.7) after an average of 320 d on feed (SD = 14). Approximately 1 mL of adipose tissue was scraped from the subcutaneous fat layer between the 10th and 11th rib of each carcass.

To provide a comparison we sampled the subcutaneous fat from the sirloins of 17 Simmental carcasses. These were slaughtered at between 9 and 14 mo, after 36 to 87 d on feed.

We heated the carcass samples, using the method described above, to extract melted fat for T_M measurements.

Slip Point Measurements

We measured slip point temperature for each standard. First, melted fat was drawn into a glass capillary tube with an internal diameter of 1 mm to form a 1-cm column. The fat solidified in the capillary tubes while stored at 4°C for at least 12 h. The position of the top of the fat column was marked on the capillary tubes, which were then attached to a thermometer marked in 1°C divisions. The thermometer was held in a beaker of water, which was heated gradually from 20 to 50°C at 1°C/min with the top of the fat column below the surface of the water. The temperature at which the fat column rose was recorded as the slip point. We measured slip points 4 times for each fat standard.

RESULTS AND DISCUSSION

Table 1 shows the Rotor-Gene Q T_M measurements of 7 aliquots of 3 different types of animal fat over 2 runs. Variation in T_M measurements on duplicated samples within a run was very low. We found that the variation between duplicates was small both within a run (average intra-assay SD of 0.08°C) and between runs (average interassay SD of 0.10°C). The assay was accurate over the range 35 to 47°C.

Currently, there are several fat T_M methods used depending on the purpose of the test. These include the slip point or open capillary method (American Oil Chemists' Society official method Cc 3b-92; AOCS, 1998), the softening point (American Oil Chemists' Society official method Cc 3-25; AOCS, 1998), the point at which the fat becomes clear (American Oil Chemists' Society official method Cc 1-25; AOCS, 1998), the dropping point (American Oil Chemists' Society official method Cc

Table 1. High-precision melting point (T_M) measurements for 3 animal fats

Sample number	T_M, °C		
	Ghee[1]	Skirt fat	Supafry[2]
Run 1			
Aliquot 1	35.42	40.90	46.32
Aliquot 2	35.40	41.07	46.25
Aliquot 3	35.43	41.17	46.35
Aliquot 4	35.52	41.40	46.28
Mean	35.44	41.14	46.30
SD	0.05	0.21	0.04
Run 2			
Aliquot 1	35.43	41.27	46.37
Aliquot 2	35.43	41.35	46.42
Aliquot 3	35.52	41.40	46.33
Mean	35.46	41.34	46.37
SD	0.05	0.07	0.05

[1]Allowrie Ghee (Fonterra, VIC, Australia).

[2]Goodman Fielder, NSW, Australia.

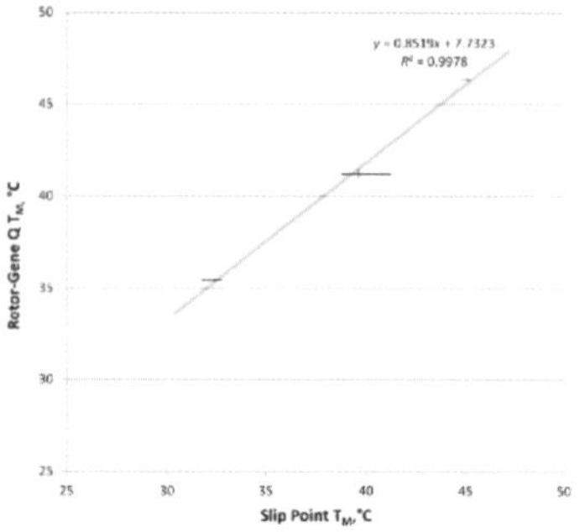

Figure 2. Correlation of melting point (T_M) measurements by slip point and Rotor-Gene Q (Qiagen, Hilden, Germany) methods.

18-80; AOCS, 1998), and differential scanning calorimetry (AOCS, 1998; Nassu and Gonçalves, 1999). Since animal fats are mixtures, their melting is a gradual process and the different definitions of the T_M in the various methods make it difficult to compare results (Nassu and Gonçalves, 1999). The slip point is the most commonly used T_M measure for fats in food production.

In Fig. 2 we compare T_M measurements using the Rotor-Gene Q assay with those using the slip point assay. The results are highly correlated but the slip point assay had more variability than the Rotor-Gene Q assay. Error bars show the large range of repeated measurements for the 4 slip point measurements compared to the small range of the 7 Rotor-Gene Q measurements. The Rotor-Gene Q T_M assay can be used in place of the standard slip point assay for various animal fats. Linear conversion is possible with Rotor-Gene Q measurements 3.3°C higher for slip points of 30 and 1.1°C higher for slip points of 45°C.

The slip point assay depends on observing fat solidified in a capillary tube rising when it melts. Slip point testing can be automated, to some extent, with purpose-designed machines. These improve the precision but not the throughput. The Melting Point M-565 (Buchi Laboratory Equipment, Flawil, Switzerland) can be used for slip point measurements of up to 3 samples at a time with a precision of 0.1°C. The Rotor-Gene Q assay described here is more reliable and accurate and is capable of testing up to 71 samples at a time.

As an example of potential applications of this technique, we measured T_M of subcutaneous fat from 67 Wagyu carcasses and 17 Simmental carcasses. We sampled 1 carcass twice, to give an indication of sampling variability, and the T_M measurements differed by less than 0.2°C. The T_M of the Wagyu samples ranged from 32.5 to 39.5°C with an average of 36.0°C and SD of 1.3°C (Fig. 3). In contrast, the Simmental samples ranged from 39.6 to 45.1°C with an average of 42.1°C and SD of 1.7°C. The phase of a fatty acid at body temperature (37°C) is known to be critical to its palatability and health effects. Of the 66 Wagyu carcasses sampled, 52 had fat that was liquid at 37°C temperature while 14 had fat that still contained some solids at this critical temperature. All of the Simmental carcasses had T_M well above body temperature.

The lower T_M of the fat samples from Wagyu is due to a combination of the softer fat from Wagyu (Pitchford et al., 2002) and the longer time on feed (Duckett et al., 1993). The variation of T_M within full-blood Wagyu raised within the same environment is largely due to genetic variation. There is potential for genetic improvements in fat composition by selecting Wagyu breeding stock that produce low T_M offspring.

We have developed a method for measuring T_M of edible fat, by application of proven technology on instruments that are available in all sophisticated laboratories.

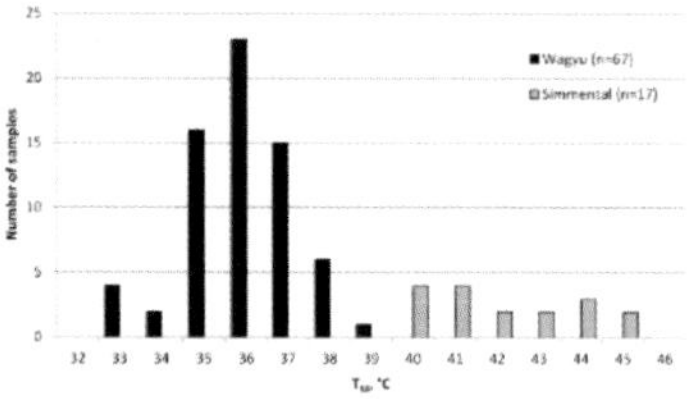

Figure 3. Melting point of the fat from Wagyu carcasses and Simmental carcasses.

Furthermore, the method described here could be encapsulated in a custom-designed apparatus with more samples and automation optimized to include the sample preparation. This method is superior to the slip T_M method and will be useful in research into improving the fat composition of meat through genetic and management strategies.

LITERATURE CITED

Adams, T. H., R. L. Walzem, D. R. Smith, S. Tseng, and S. B. Smith. 2010. Hamburger high in total, saturated and trans-fatty acids decreases HDL cholesterol and LDL particle diameter, and increases TAG, in mildly hypercholesterolaemic men. Br. J. Nutr. 103:91–98.

American Oil Chemists' Society (AOCS). 1998. Official methods and recommended practices of the AOCS. 3rd ed. AOCS Press, Champaign, IL.

Daley, C. A., A. Abbott, P. S. Doyle, G. A. Nader, and S. Larson. 2010. A review of fatty acid profiles and antioxidant content in grass-fed and grain-fed beef. Nutr. J. 9:10.

De Smet, S., K. Raes, and D. Demeyer. 2004. Meat fatty acid composition as affected by fatness and genetic factors: A review. Anim. Res. 53:81–98.

Duckett, S. K., D. G. Wagner, L. D. Yates, H. G. Dolezal, and S. G. May. 1993. Effects of time on feed on beef nutrient composition. J. Anim. Sci. 71:2079–2088.

Garmyn, A. J., G. G. Hilton, R. G. Mateescu, J. B. Morgan, J. M. Reecy, R. G. Tait, D. C. Beitz, Q. Duan, J. P. Schoonmaker, M. S. Mayes, M. E. Drewnoski, Q. Liu, and D. L. VanOverbeke. 2011. Estimation of relationships between mineral concentration and fatty acid composition of longissimus muscle and beef palatability traits. J. Anim. Sci. 89:2849–2858.

Gilmore, L. A., R. L. Walzem, S. F. Crouse, D. R. Smith, T. H. Adams, V. Vaidyanathan, X. Cao, and S. B. Smith. 2011. Consumption of high-oleic acid ground beef increases HDL-cholesterol concentration but both high- and low-oleic acid ground beef decrease HDL particle diameter in normocholesterolemic men 1, 2. J. Nutr. 141:1188–1194.

Knothe, G., and R. O. Dunn. 2009. A comprehensive evaluation of the melting points of fatty acids and esters determined by differential scanning calorimetry. J. Am. Oil Chem. Soc. 86:843–856.

Mensink, R. P., P. L. Zock, A. D. Kester, and M. B. Katan. 2003. Effects of dietary fatty acids and carbohydrates on the ratio of serum total to HDL cholesterol and on serum lipids and apolipoproteins: A meta-analysis of 60 controlled trials. Am. J. Clin. Nutr. 77:1146–1155.

Micha, R., and D. Mozaffarian. 2010. Saturated fat and cardiometabolic risk factors, coronary heart disease, stroke, and diabetes: A fresh look at the evidence. Lipids 45:893–905.

Nassu, R. T., and L. A. G. Gonçalves. 1999. Determination of melting point of vegetable oils and fats by differential scanning calorimetry (DSC) technique. Grasas Aceites 50:16–22.

Paniagua, J. A., A. G. de la Sacristana, E. Sánchez, I. Romero, A. Vidal-Puig, F. J. Berral, A. Escribano, M. J. Moyano, P. Peréz-Martinez, J. López-Miranda, and F. Pérez-Jiménez. 2007. A MUFA-rich diet improves posprandial glucose, lipid and GLP-1 responses in insulin-resistant subjects. J. Am. Coll. Nutr. 26:434–444.

Pitchford, W. S., M. P. B. Deland, B. D. Siebert, A. E. O. Malau-Aduli, and C. D. K. Bottema. 2002. Genetic variation in fatness and fatty acid composition of crossbred cattle. J. Anim. Sci. 80:2825–2832.

Uemoto, Y., T. Abe, N. Tameoka, H. Hasebe, K. Inoue, H. Nakajima, N. Shoji, M. Kobayashi, and E. Kobayashi. 2010. Whole-genome association study for fatty acid composition of oleic acid in Japanese Black cattle. Anim. Genet. 42:141–148.

Analysis of Haplotype Sequences

Chapter 12

Analysis of Haplotype Sequences

Sally S. Lloyd, Edward J. Steele and Roger L. Dawkins

Additional information is available at the end of the chapter

http://dx.doi.org/10.5772/61794

Abstract

In this era of whole-genome, next-generation sequencing, it is important to have a clear understanding of the concept of "haplotype". We show here that most of the important regions of the genome can be described in terms of polymorphic frozen blocks (PFB). At each PFB, there are numerous, even hundreds, of alternative ancestral haplotypes. Haplotypes, not genes, can be regarded as the principal unit of inheritance. We illustrate how sequence data can be analysed to reveal and define these ancestral haplotypes.

Keywords: Ancestral haplotypes, Polymorphic frozen blocks, Genomic evolution

Next Generation Sequencing - Advances, Applications and Challenges

Edited by Jerzy K Kulski

Analysis of Haplotype Sequences 351
http://dx.doi.org/10.5772/61794

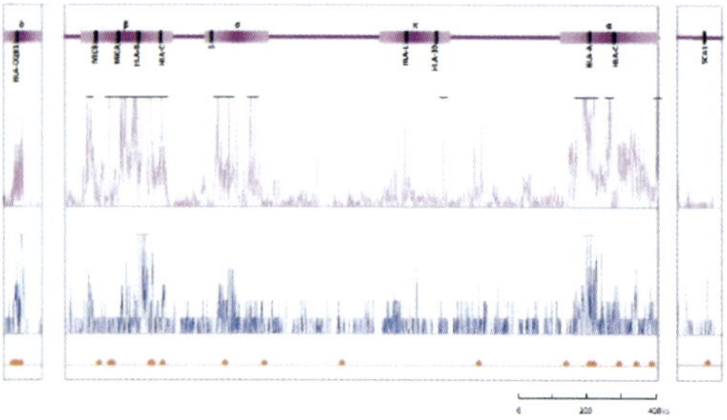

Adapted from ref. [26].

Figure 3. Sequence diversity is packaged as polymorphic frozen blocks (PFB). SNPs and indel occur in similar locations within PFB. (a) The SNP profile after removing indels. Peaks higher than 20 SNPs per 1000 nucleotides are truncated. (b) The location of indels. Peaks higher than six indels per 1000 nucleotides are truncated. (c) The position of indels greater than 100 nucleotides.

Soma to germline feedback is implied by the extreme polymorphism at IGHV relative to MHC

Bioessays 37:557-569

Prospects & Overviews

Problems & Paradigms

Soma-to-germline feedback is implied by the extreme polymorphism at IGHV relative to MHC

The manifest polymorphism of the MHC appears greatly exceeded at Immunoglobulin loci, suggesting antigen-selected somatic V mutants penetrate Weismann's Barrier

Edward J. Steele and Sally S. Lloyd*

Soma-to-germline feedback is forbidden under the neo-Darwinian paradigm. Nevertheless, there is a growing realization it occurs frequently in immunoglobulin (Ig) variable (V) region genes. This is a surprising development. It arises from a most unlikely source in light of the exposure of co-author EJS to the haplotype data of RL Dawkins and others on the polymorphism of the Major Histocompatibility Complex, which is generally assumed to be the most polymorphic region in the genome (spanning ~4 Mb). The comparison between the magnitude of MHC polymorphism with estimates for the human heavy chain immunoglobulin V locus (spanning ~1 Mb), suggests IGHV could be many orders of magnitude more polymorphic than the MHC. This conclusion needs airing in the literature as it implies generational churn and soma-to-germline gene feedback. Pedigree-based experimental strategies to resolve the IGHV issue are outlined.

Keywords:
IGHV; MHC ancestral haplotypes; polymorphic frozen blocks; somatic hypermutation; soma-to-germline inheritance

Additional supporting information may be found in the online version of this article at the publisher's web-site

Introduction

There is a dawning realization that germline regions encoding immunoglobulin (Ig) variable (V) genes harbor extreme DNA sequence polymorphism. This paper compares IGHV

DOI 10.1002/bies.201400213

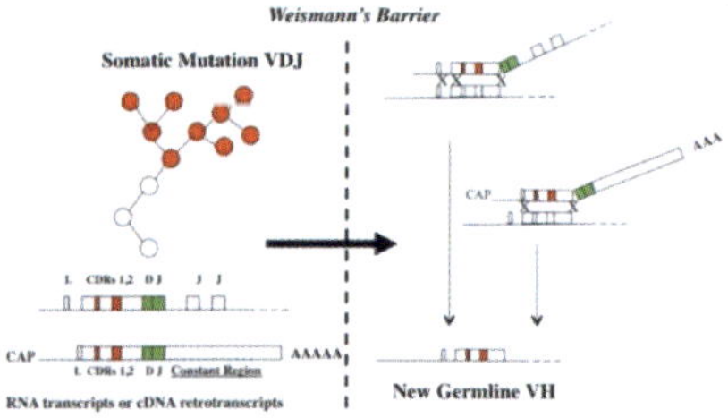

Somatic mutations in CDR1 and CDR2 (red) or CDR 3 regions (green) of VDJ genes are clonally selected in B lymphocytes. RNA or cDNA copies penetrate Weismann's Barrier for targeted homologous integration (x, gene conversion) of the V portions (harboring CDR1, 2) into similar germline VH

In The Press

Wagyu's Lower Melting Point a Heart Tick

NEW RESEARCH FINDINGS CONFERENCE NEWS

WAGYU'S LOWER MELTING POINT **A HEART TICK**

Long fed Wagyu beef can legitimately claim to be healthier than most other beef according to research scientist Dr Sally Lloyd who was responding to a question at the World Wagyu Conference.

Fat in Wagyu beef, especially long fed, had a melting point of at least 5°C below the average of other breeds - this brings the melting point to below body temperature meaning the fat melts in the mouth for a better taste sensation and is 'better for you' according to Dr Lloyd, Research Director at the West Australian based CY O'Connor (CYO) Foundation.

"A lower fat melting point delivers healthier beef with less saturated fats," said Dr Lloyd who likened Wagyu fat to olive oil with better cholesterol and blood lipids resulting in better cardiovascular health.

Dr Lloyd has also developed a new test for fat Tm (melting point) which is automated and more accurate and less cumbersome and labour intensive than the traditional 'slip point' system. This new Rotorgene Q Inversion point system is 'patent pending' with the CYO Foundation.

"Fat melting point (Tm) is the simplest measure of the healthiness of fats and beef with lower melting point fat is significantly healthier", she said.

Long fed Wagyu beef fat has a Tm of 33°C while short fed European breeds produce fat with a melting point of 44°C. Wagyu's 33°C falls in the same area as cocoa butter and palm oil while the 44°C sits between the melting points of butter and carbohydrates.

Randomized, controlled studies in the USA evaluated individuals who consumed ground beef formulated from long grain-fed steers for five weeks (five patties a week) compared to the consumption of regular ground beef - lower in oleic acid. HDL (good) cholesterol increased significantly in normocholesterolemic men and postmenopausal women fed the high oleic acid ground beef. In these studies the men consumed ground beef containing 24% fat and the women consumed ground beef containing 20% fat.

"We already know consumers prefer to eat beef with marbling", said Dr Lloyd who referred to Australia's MSA research and meat science data from the USA.

For the past two years Dr Lloyd has recorded the Tm of the entries in the Wagyu Branded Beef Competition and has found that Fullblood Wagyu have a lower Tm than Crossbred Wagyu.

Dr Lloyd urged Wagyu breeders to use genetic testing and selection to further improve (lower) the Tm of Wagyu beef because beef with lower Tm is healthier and tastes better. She said Tm is heritable and fast progress could be made through genetic testing with breed differences showing up some unique haplotypes in different frequencies. Wagyu specific haplotypes show higher and lower fat Tm.

>>

ABOVE RIGHT

Dr Sally Lloyd told the World Wagyu Conference that low melting point fat found in Wagyu beef is a 'healthy' fat

AUSTRALIAN WAGYU UPDATE www.wagyu.org.au 11

BEEF WITH LOWER MELTING POINT FAT IS SIGNIFICANTLY HEALTHIER

Change in total to HDL cholesterol ratio expected from changing 10% of calorie intake from average US dietary fat to a variety of other foods, based on a figure from *Mensink 2003.*

The expected change in total to HDL cholesterol ration for high and low Tm beef are calculated from the fatty acid compositions taken from *Smith 2006*, according to the formula given in *Mensink 2003.*

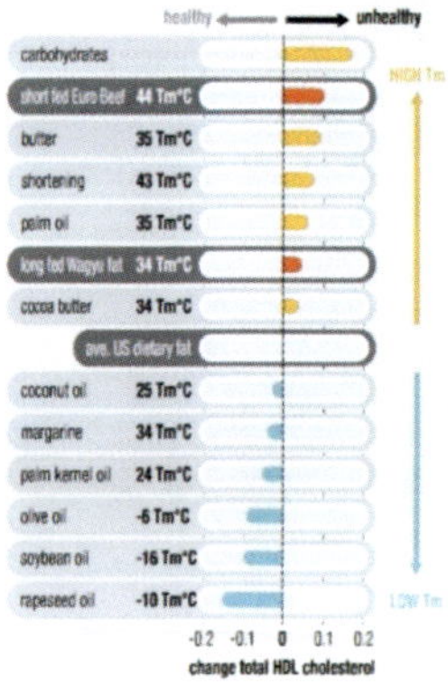

PASTURE FEEDING CAN INCREASE Tm BUT NOT FOR ALL ANIMALS

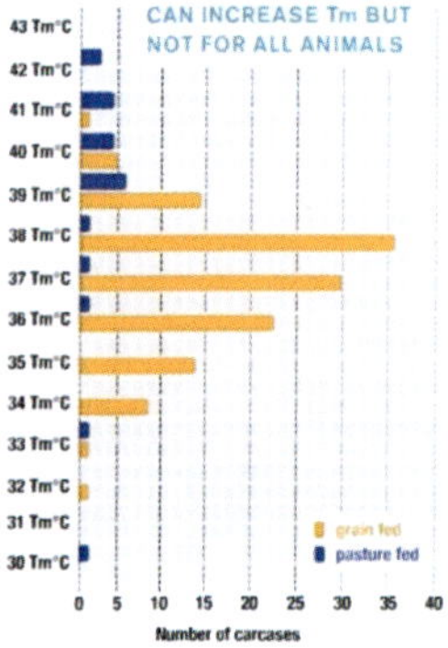

Wagyu's lower melting point a heart tick

Continued from page 11

Dr Lloyd compared the fat Tm of pasture fed and grain fed carcases and while pasture feeding tended to increase Tm, it was not the case for all animals.

The Australian research at the CY O'Connor Foundation appears to endorse extensive research at Texas A&M (USA) where it was found that heavily marbled long-grainfed Wagyu beef contains 'extraordinary' amounts of the monounsaturated 'healthy' oleic fatty acid.

Oleic acid is abundant in olive oil and its presence in beef increases with marbling content and days on feed.

"We need fat in beef to improve the eating experience," said Dr Stephen Smith, a highly regarded research scientist and Regents Professor in animal science at Texas A&M.

Smith teamed with Dr Brad Johnson of the Texas Tech University to co-author a paper, 'Marbling: Management of cattle to maximise the deposition of intramuscular adipose tissue'. The paper addresses the biology and biochemistry of beef marbling and its effects on production systems, carcase and fat quality.

"In our research, we examined young cattle just before they marbled and were primarily looking at genes related to fatty acid composition," he said.

"As cattle fatten and put down marbling, the fat becomes healthier because there is a replacement of saturated fats with oleic acid".

The research also attempted to discover the cellular processes that regulate this very natural increase in oleic acid in beef.

The researchers looked at gene expression associated with fat development. In general terms in transitioning from pasture or grass feeding to feedlot feeding there is a profound increase in genes associated with fat development and making more oleic acid.

"You can barely detect expression of genes related to marbling and fat composition in cattle on pasture, but much more so when cattle are fed grain," said Dr Smith.

"The longer they are on feed the more oleic acid they deposit. If you take Wagyu or Hanwoo, which are grain fed up to 30 months of age they have an extraordinary amount of marbling and oleic acid. Wagyu and Hanwoo beef marbling fat is very soft which provides a juicy mouth feel."

The study indicates that marbled beef increased high density lipoprotein (HDL) cholesterol - the good cholesterol - in men and women and the relationship between fat and overall palatability "underscores the importance of grain feeding and intramuscular lipid in beef quality".

As fat increases, it is accompanied by a decrease in the proportion of the 'bad' saturated and trans-fatty acids with a corresponding increase in the 'good' oleic and other monounsaturated fatty acids.

"The more cattle fatten, (the more) they put down more marbling and the more healthful beef is," Smith said.

"We hope to convince everyone in the beef production chain, all the way from producers to retailers, that healthy fat in beef not only improves flavor, but you can modify the animal naturally so that beef contains more oleic acid. This provides a very palatable product that, even though it contains relatively high levels of fat, is not going to have a negative effect on cholesterol metabolism in humans," said Smith. ■

MORE INFORMATION

about Marbling: management of cattle to maximise the deposition of intramuscular adipose tissue can be downloaded from

http://bit.ly/1EwH8x6

Beef from paddock to plate

CHOLESTEROL CUTTER

Beef from paddock to plate

Rob Broadfield

Fresh produce: Professor Roger Dawkins with Wagyu beef cattle in a paddock behind the C. Y. O'Conner Village Pub in Piara Waters. Picture: Nic Ellis

Give a world-renowned geneticist a suburban tavern, some smuggled genes and a rare super-breed of Wagyu cattle and something extraordinary is bound to happen.

In the case of Professor Roger Dawkins, it means Australia's only tavern with a Wagyu herd out back and a groundbreaking research program that may make our beef tastier and healthier.

Professor Dawkins, credited with breakthroughs in the science of immunogenetics in Australia and for developing some of the world's first diagnostic tests for AIDS, has a passion for cattle, good beef and genetics.

His pub, the C.Y. O'Connor Tavern at Piara Waters, has to be one of the most unusual scientific test beds for high-end genetic research.

"It's all about health," Professor Dawkins said as we sat down to a meal of aged Wagyu this week.

"Levels of heart disease are rocketing in the Western world and the healthy fat of the Wagyu breeds may well provide an answer for meat lovers."

The research project aims to breed the healthy, low melting point fat and marbling that have made the Wagyu breeds famous into mainstream breeds of cattle.

"Wagyu fat has a melting point of below 37C, whereas the more common breeds — Angus and Murray Grey for instance — have a much higher melting point at 42C and above," Professor Dawkins said.

"This means Wagyu's good fat remains soluble at human body temperature and does not contribute to high cholesterol or the build-up of arterial plaque."

The C.Y. O'Connor Research Foundation, which has its laboratories on the family farm behind the tavern and is named in honour of Professor Dawkins' great-great-grandfather, Kalgoorlie pipeline builder Charles Yelverton O'Connor, has produced modelling suggesting that eating a Wagyu steak reduces cholesterol.

Wagyu fat has a high concentration of oleic acid, which is found most commonly in olive oil, and is well-known to have a positive effect on cholesterol in humans. Professor Dawkins said eating Wagyu could reduce the nation's reliance on statins, the expensive cholesterol-beating drugs widely prescribed in Australia.

Tasty: John Dawkins

His research has led his team to the lesser-known Red Wagyu, a similar Japanese breed to the Black Wagyu but with superior genetics that allow it to marble on a diet of grass and with a greater tolerance for subtropical climates.

Black Wagyu usually do not produce the desirable marbling on grass and are mostly raised on grain.

"The tropical breeds like Santa Gertrudis and Brahman are not good eating in terms of prime beef, so to be able to improve Red Wagyu genetic characteristics in tropical breeds, potentially opens up a massive new front for beef exports from northern Australia," Professor Dawkins said.

He said Australia has the best gene pool for the red breed, known as Aguashi in Japan.

"American breeders and researchers are constantly seeking Australian genetics to improve their herds," he said.

"We came by these Red Wagyu genes in the 1980s when, somehow, they arrived in Australia. It is illegal to export Wagyu cattle or semen from Japan. They hold on to it very tightly."

But the proof is in the pudding, as they say, and the Wagyu steaks at the C.Y. O'Connor Tavern are equal to any Wagyu steaks in any of the finest restaurants in Australia.

The Black Wagyu is still superior eating but Professor Dawkins expects Aguashi meat to improve as the breeding program continues.

Professor Dawkins and his son John, who runs the tavern and cattle operations, are soon to present their Wagyu — under their Village Green brand — to top Perth chefs.

Travel Reports

World Wagyu Conference 2015

Dr. Sally Lloyd

In May 2015 I travelled to Rockhampton on invitation from the Wagyu Association to present CYO's latest research to the Inaugural World Wagyu Conference.

My talk was well received. Producers were keen to include health benefits of lower Tm fat in the marketing of Wagyu. Several breeders expressed interest in testing their herds to identify low Tm genetics within their herds.

In particular Gerard Davis from AACo, Nick Sher from Sher Wagyu and Gerard Hickey from First Light Foods in New Zealand.

The New Zealand Government has part funded a $25million research program, with First light foods into raising F1 Wagyu on pasture, as a way of improving the quality of meat. Their lab does meat quality and taste testing and would be interested in adding a melting temperature test to their capabilities.

American Wagyu Association

Professor Roger L Dawkins

I attended the American Wagyu Association annual meeting held in Ocala, Florida October 21 to 23, 2015.

My talk appeared to create interest as judged by subsequent discussions and by invitations to join ongoing studies. Many of the issues were similar to those debated by local breeders.

Most interesting was the need to improve testing for food quality and degrees of infusion. I was impressed by the productive relationship between the independent testing laboratory at Geneseek and the registry.

Following visits to several sites in Texas, including large scale producers, I attended the American Akaushi Association meeting help in Albuquerque, New Mexico October 29 to 31, 2015. There was an emphasis on genetic testing especially for recessive disorders which appear to be largely solved.

On October 26 to 28, 2015, I had the pleasure of discussing patents and strategies with the Foundation's patent attorneys at Sheridan Ross in Denver.

Along the way, I was able to sample excellent steaks – although none seemed as good as our own.

Scientific Retreat 2015

4th-5th March Albany, WA

The CYO Foundation and Hybla joined with Irongate Wagyu at Kalgan River, Albany (Mr Peter Gilmour and family) for the Albany Retreat.

Wednesday 4th March- Irongate Wagyu Open Day

8.45am	**REGISTRATION**
9.00am	**Welcome** Welcome by AWA and host Irongate Wagyu. Introduction of the speakers and program for the day.
9.10am	**Understanding the Wagyu value chain** Including calving ease, marbling genetics, docility, longevity and carcass value.
9.30am	**Wagyu Genetic Analysis and how it works** An interactive discussion on the value of genetics, traits of importance, selection methods and an introduction to Estimated Breeding Values (EBVs). Recognises Wagyu breeders make wide use of AI and ET as well as natural sires and dams.
9.50am	**Research carcass EBVs and the new Wagyu Terminal and Self-replacing Indexes** The Wagyu Collaborative Genetics Research Project has revolutionised the analysis and selection of Wagyu genetics. New EBVs and indexes will be explained.
10.10am	**Melting point of Wagyu intramuscular fat and healthy fats**
10.40am	**MORNING TEA**
11.00am	**Registering and on-going performance recording of breeding and slaughter animals** All Wagyu breeding members need to register and performance record their breeding herd to maximise herd value, breeding potential and profitability.
11.30am	**The value of historic feedlot and carcase data and how to load it** Historic feedlot weights and Aus-meat data can be a rich source of starting data.
11.50am	**Using EBVs and indexes for selection and purchase of breeding animals** Using EBVs and other genetic technologies in Wagyu bull and female selection including the use of the AWA online Wagyu animal search database.
1.00pm	**LUNCH**
2.00pm	**Utilising Matesel and Geneprob** MateSel enables computer optimised mating using EBVs/indexes. GeneProb predicts the probability of genetic conditions in animals. Learn how to use these valuable tools.
2.30pm	**Wagyu Collaborative Genetics Research Project - Stage 2 update** The Research Project is now moving into its second stage across the next 5 years. Hear how this will affect your breeding operations.
2.50pm	**Marketing Wagyu Fullbloods, Crossbred Wagyu and seedstock opportunities** Wagyu offers many opportunities for improved profitability, but make sure you get the marketing right.
3.30pm	**Workshop conclusion**
3:30 - 5:00pm	**Tour of the Irongate Wagyu property and its operations**

A well attended open field day at Irongate Wagyu with presentations from Australian Wagyu Association (AWA) including CEO Craham Truscott,and Wagyu Technical Officer and ABRI CEO Alex McDonald. Together with Peter Gilmour they covered the following topics: Understanding the Wagyi value chain; Wagyu Genetic Analysis and how it works; Research carcass EBVs and the new Wagyu Terminal and Self-replacing indexes; Registering and on-going performance recording of breeding and slaughter animals; The value of historic feedlot and cascass data and how to load it; Using EBVs and indexes for selection and purchase of breeding animals; Utilising Matesel and Geneprob; Wagyu Collaborative Genetics Research Project – Stage 2 update. And a final wrap up presentation by Peter Gilmour "Marketing Wagyu Fullbloods, Crossbred Wagyu and seed stock opportunities"

Dr Sally Lloyd presented for CYO on "Melting point of Wagyu intramuscular fat and healthy fats" which was very well received.

A gourmet Wagyu lunch and special tastings was then graciously provided by Peter Gilmour and family.

Later Mark James , Irongate Herd Manager took the attendees on a tour of the Irongate Wagyu property and its operations.

In the evening. All attendees were invited to the mezzanine room at Hybla Bar and Grill Middleton Beach, Albany where Dr Sally Lloyd and CYO Foundation and Hybla staff conducted a blind beef tasting completion as well as a much enjoyed wine tasting. Mr Gary Park of Hamika Grazing Co. correctly identified all samples and their quality.

Thursday 5th March – Hybla

SYMPOSIUM 1: 9:00 – 12:00 Traceability and Food Security

The CYO Foundation conducted a special Retreat at Hybla attended by Graham Truscott, Alex McDonald, Peter Gilmour, Roger Dawkins, Sally Lloyd, Ted Steele, John Dawkins, Michael Lloyd, Tony Lloyd and Roger Stubbs.

Topics included: Professor Roger Dawkins – "The Quantal Structure of the Genome- Critique of SNP and Genomic EBV approaches to live stock selection "; Dr Sally Lloyd – "MRIP Haplotypes and Healthy Beef" ; Mr Tony Lloyd (for Bob Butcher) " Foetal Bovine Serum "; Michael Lloyd " Raising Sheep in Salt Bush Country" and Dr Ted Steele " A new paradigm in Science publishing: The Freedom of Print on Demand (POD)."

LUNCH SYMPOSIUM : 12:00-14:00 by special invitation
Speakers: Jeremy Stewart, Micheal Roberts, Roger Dawkins
The discussion focused on how to ensure the esplanade site contributes to all year all weather attractiveness of Ellen Cove and supports research and development in health, evolution climate change and food.

9th-10th April: Individuality & Identification CYO'Connor Village

The Keynote Speaker & leader of a Practical Workshop was Sergeant Graham Byard, Research, Development and Accreditation Manager, Fingerprint Bureau, WA Police Forensics Division.

Symposium 1: Introduction to Individuality

BACD for Species Identification *RL Dawkins*
Nose Print ID in cattle *J Valenzuela*
Forensic Fingerprinting -1 *G Byard*
Individuality and Recognition-RL Dawkins

Symposium 2: Practical Workshop – Forensic Fingerprinting

Collection of Fingerprints *G Byard*
Bone Marrow Transplantation, Graft v Host- *L Smythe*
Preservation, Storage, Interpretation and Reporting of Fingerprints *G Byard*
Human Mutation rates *EJ Steele*

Symposium 3 : Practical Workshop – Forensic Fingerprinting

Scleroderma and changes in fingerprints in autoimmunity *S Lester/L Smythe*
Forensic Fingerprinting-Future Needs *G Byard*
Multifunctional Polymorphic Controller *RL Dawkins*

Symposium 4: Selective Breeding in Agriculture

Principles Livestock Selection *RL Dawkins*
Haplotype Analysis Human Olfactory Receptor Clusters*EJ Steele*
Challenges in the Dairy Industry *R Giura*

Symposium 5: Selective Breeding in Agriculture – Practical Aspects

Logic of Next mating and AI Program *JR Dawkins*
Extraction DNA and Fat in a single Step *A Wijewickrama*
Livestock Selection Based on Beef Fat Tm *SS Lloyd*
The Australian Experience with Wagyu *D Bayard*
Diffrences between Wagyu Bulls & MRIP Haplotypes *P Gilmour*
Red Wagyu (Akaushi) Past and Future *B Reed*
Deficiences in MSA Grading *G Park*
Kikuyu Grass *R Giura/RL Dawkins*

Symposium 7: CYO's Strategic Position

Value of PFB, Recombination suppression in selective breeding *A Lloyd*
The new PEEL Research Centre *R Stubbs*

Saturday 11th April : Tour of Roebrooks Farm , Cattle viewing & flushing AI/ET Program (Dom Bayard)

8th-10th October: Towards Low Melting Point Fat in Wagyu, CYO'Connor ERADE Village, Canning Vale

Thursday 8th October

C Y O'Connor ERADE Village, CYO Board Room

Symposium 1: The CYO Research Program

Directions for Wagyu Research *Roger Dawkins*
Rumen Pangenome Project and Significance for Northern Beef Industry *Phil Vercoe*
Tm as the Best Measure of Beef Quality *Sally S Lloyd*
Genetic associations with low Tm in Pure Wagyu *Sue Lester/Sally Lloyd/Roger Dawkins*

Symposium 2: The Production of Reliable Livestock Meats

F1 strategy in pig, chicken and grain industries *Anthony Lloyd*
Selecting Best Dam in Cross Breeding at Strathdale Wagyu *Darren Hamblin*
Selecting Best Wagyu Sire in Cross Breeding (Brahman, Santa Gertrudis etc.) *Peter Gilmour/Mark James*
Low Tm at ≤100 days on feed? *Sally Lloyd*

Symposium 3: Wagyu Ventures in the Kimberley

Wagyu Infusion Plans on Pardoo Station *Alan Peggs*
Anna Plains Station nr Roebuck south of Broome (Brahman) *David Stoate*
Akaushi (Red Wagyu) at Mazda Wagyu *Bob Reed*
Robbins Island Wagyu (soft fat gene tests, marbling fineness) *Keith Hammond*

Symposium 4: Student Presentations

Metabolic Indicators in Wagyu Selection *Samantha Connelly*
Comparison Between US and Australian Cattle Production Systems *Alex Hammond*
PCR Haplotyping -Tracking Animals from Paddock to Plate *Nicole Warner*

Symposium 5: The Wrap – CYO's Search for Genetic Factors Influencing Low Tm

A Summing Up& Group Discussion (Sire Selection and Low Tm, C19 Haplotypes, Interactions and Low Tm Associations)
Leader: Roger L Dawkins
Animal Science at UWA *Phil Vercoe*

CYO Foundation Dinner including Fund Raising Auction

Friday 9th October

C Y O'Connor ERADE Village, CYO Board Room

Symposium 6: CYO Associates and Student Presentations

Interactions Tm, chiller temperature, BMS/camera *Sally Lloyd*
Practicalities of Camera at the Abattoir *Peter Gilmour*
CYO Cattle Database: ID, Sire Families, Haplotypes, Tm, Polling *EJ Steele*
Milk feeding of Weaners *Rob Giura*
Strategy for Developing New Markers for Meat Quality
Jose Valenzuela/ Sally Lloyd
Sexed Semen: Separating X and Y carrying Spermatozoa
Catherine Skinner
New Generation Tm – Equipment, Samples, Collaborations
Sally S Lloyd
A Lesson in *cis, trans*, and epistatic Interactions: Effects on Crossbreeding and Relevance to EBVs *Sue A Lester*

Tour of Melaleuka Stud & Nth Dandalup

Information on Participants

Professor Roger L Dawkins
Chair CYO Foundation/ Melaleuka Stud
Mr John R Dawkins *Principal, Melaleuka Stud, Nth Dandalup*
Mr Anthony Lloyd *Vice Chairman, CYO Foundation*
Mr Roger Stubbs *Board Member, CYO Foundation*
Dr Sally S Lloyd *Research Director, CYO Foundation*
Dr Edward J Steele
Honorary Research Associate, CYO Foundation
Dr Jose Valenzuela
Research Student CYO Foundation/Melaleuka Stud
Ms Nicole Warner *Technical Trainee, CYO Foundation*
Ms Catherine Skinner *Technical Trainee, CYO Foundation*

Ms Susan Lester *Rheumatology, Queen Elizabeth Hospital, SA*
Mr Peter Gilmour *Principal, Irongate Wagyu, Albany, WA*
Mr Mark James *Herd Manager, Irongate Wagyu, Albany, WA*
Mr Darren Hamblin
Principal, Strathdale Wagyu, Middlemount, QLD
Ms Samantha Connelly
PhD Scholar (QLD) Strathdale Wagyu, Middlemount, QLD
Mr Tim Hyde
Managing Director, Advanced Fertigation Systems, Carnarvon, WA
Mr Brett Blanchett
General Manager, Liveringa Station Beef, WA
Mr David Stoate
Principal, Anna Plains Station, Roebuck, WA
Mr Alex Hammond *Robbins Island Wagyu, Tasmania*
Professor Phil Vercoe *Professor Animal Science, UWA*

Summary of the CYO Wagyu Retreat, October 2015

This was probably the largest and most successful Retreat held in a long time.

On Thursday 8th and Friday 9th October 2015 the CYO Foundation ran a Retreat focused on Wagyu. Most talks were on Thursday followed by a charity fund raising dinner, whilst on Friday interstate visitors were taken on a tour of the Melaleuka Stud at Roebrooks Farm and the North Dandalup properties. The goal of the meeting was to develop selective breeding strategies to produce Wagyu or crossbred Wagyu animals which reliably marble and produce low Tm ($\leq 37^{o}C$) beef for the very minimum days on feed. The other main goal was to discuss the Wagyu infusion plans and progress made for the northern WA regions (Kimberly).

Professor Roger Dawkins welcomed the group and then expounded on the value of CYO's haplotype approach to the genetic selection of livestock. Amongst other matters he focused on the key industry issue : how we can " develop by selective breeding Wagyu or crossbred Wagyu animals which reliably fine marble and produce low Tm (≤37°C) beef for the very minimum days on feed (<300 days)." He presented CYO preliminary data which strongly suggests a significant effect in both Wagyu and Wagyu X of animals homozygous for the TCAP20 allele displaying lowered Tm of their beef fat at less than 150 days on feed. Obviously this result will be pursued to check on its validity (Figure 1) – but the economic aim clearly is to progressively select for animals, by whatever genetic indicator, in the bottom left quadrant **Low Tm, Low DOF** .

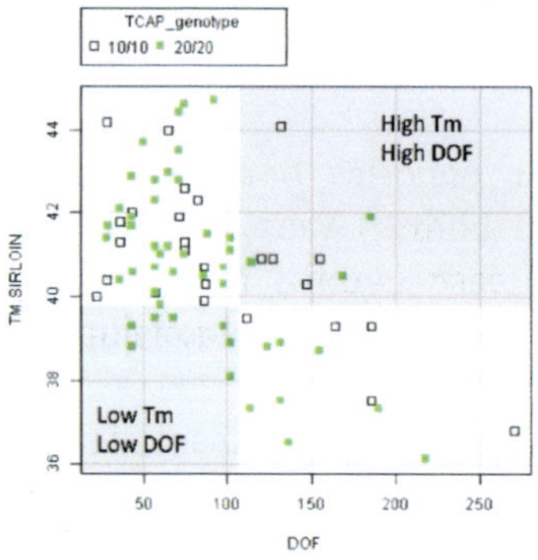

	TCAP 10	TCAP 20	Total
High Tm High DOF	5	3	8
Low Tm Low DOF	0	7	7
Total	5	10	15

P value 0.026

Genetic marker TCAP 20,20 shows a steeper decrease in Tm with DOF

TCAP_genotypes in homozygous state for Tm of beef fat versus days on feed Wagyu, Wagyu X

Towards the end of his presentation **Dawkins** reminded the group that the traditional selective breeding strategy used in developing the poll Dorset sheep in SA by his uncle Will Dawkins and others (in the 1930s-1950s) involved cycles of intense inbreeding followed by cleansing. This approach is very successful but slow and steady and maybe the ultimate preferred approach for Wagyu – an unstated qualification would be an adherence to Takeda-type rotational breeding system balancing Tm v marbling v frame size in maintaining a good herd (e.g. Irongate Wagyu , below) .

Professor Phil Vercoe then presented a summary of the research undertaken at the Institute of Agriculture, University of Western Australia. In outlining the International Ruminant Pangenome Project the distillation is that achieving transformational change depends on tackling systems and optimising the interaction between Genetics, Environment and Management or "the Right animals, in the right place, managed in the right way". The aim of the Pangenome project is to 'maximize our impact on methane emissions' and feed efficiency. Such a systems approach he asserted would be crucial for the success of Wagyu infusions in the Kimberly.

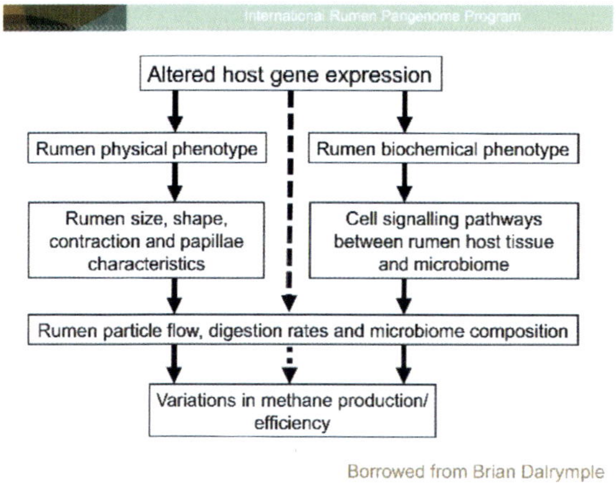

The key Wagyu industry participants were :

• Peter Gilmour and Mark James of Irongate Wagyu (Albany, WA)

• Darren Hamblin of Strathdale Wagyu (Middlemount, QLD) and PhD student Samantha Connelly

• Keith Hammond (and son Alex) of Robbins Island Wagyu (Tasmania)

• Bob and Jan Reed, Mazda Wagyu International, Akaushi genetics, (Victoria)

• Alan Peggs, Pastoral Agent representing Pardoo Station (Kimberly, WA)

• David Stoate, Anna Plains Station (Kimberley, WA)

• Andrew Rado, Stone Axe Pastoral, WA.

Darren Hamblin presented a very interesting story on how he started out with a Brahman stud but the inability to realise breeding gains switched to Wagyu Crossbreds in 2001 and used Wagyu semen with sourced Wagyu X females (Shorthorn) to expand faster. This led to six properties extending from Middlemount (242 km inland from Mackay and Rockhampton) through coastal properties between Nebo and Sarina and one in the south on the Darling Downs. The emphasis has been on producing quality Wagyu X calves and inseminating 2500 females per annum. A key to the Hamblin system in selecting the best dam in cross breeding is the Progeny Score (PS) or in-house "EBV", calculated from the carcass data results from all sire/dams progeny – continual improvement based on marbling and weight gain scores. The database uses a practical "Stockbooks" system with the computer terminal at each crush on each property run by the property manager with data entry organised by a fulltime employee. It is a web based synchronised database system which is accessible by all employees. The extent of the organisation of the development of the Hamblin herd was manifest. The plan for metabolic monitoring was outlined by **Samantha Connelly.** Indeed working with University of Sydney collaborators she we will carry out an ambitious program of analysis of samples from the Hamblin herd for key metabolic intermediates via Nuclear Magnetic Resonance (NMR) , and find the best molecular correlation signatures with marbling, weight, tenderness and EMA.

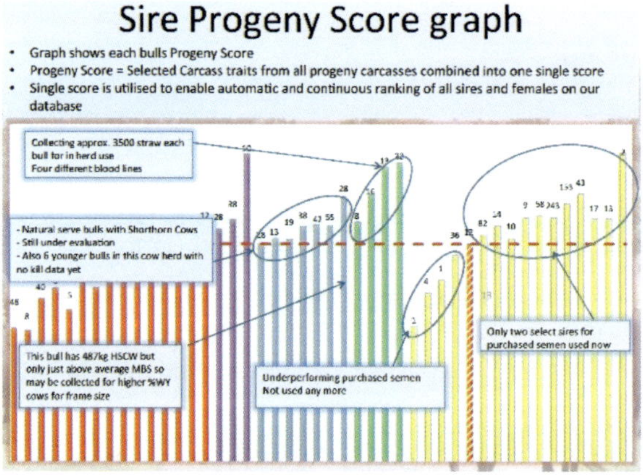

The red dotted line represents the PS performance of Michifuku as a reference in the Hamblin herd

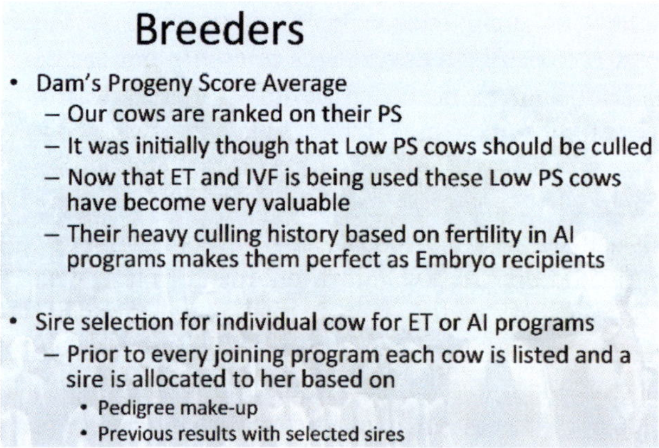

Peter Gilmour and **Mark James** discussed theIrongate Wagyu outcrossing strategy for selecting the best Wagyu sires, as well as how marbling scores are determined by experts (in Japan v Australia) and by camera. At Irongate Takeda rotational breeding balancing frame , marbling, milk, calf raising ability is their desired goal. **Peter Gilmour** also took us through the Australian Wagyu Associations Wagyu genetic Analysis (BREEDPLAN) , EBVs to predict genetic performance and Indices to rank animals. Constant feed back of animal carcase data is the key to improvement.

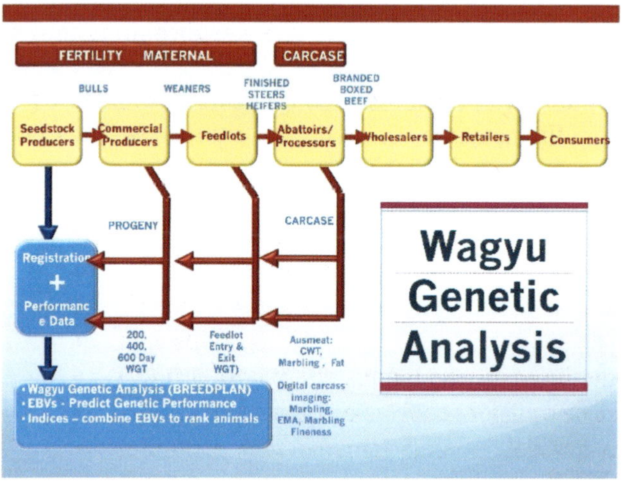

The progeny performance of a number of key Irongate sires

was covered at some length (Irongate Itoshigenami IGWFB0006, Nomby Ichuruno IGWFC0016,Itoshigenami H13 IGWFH0013) – in particular Itoshigenami IGWFB0006 for n=33 calves approx. 300 DOF average BMS (marbling) was 8.6 for an avg CWT of 430.95kg, for final average wt 700.85kg. His recommendation for selecting the best Wagyu sires for crossbreeding to Brahman , Santa Gertrudis , Droughtmaster , Brangus , Shorthorn , Angus was to "chase the marbling performance". He like **Professor Dawkins** holds high hopes for developing a genetic test for marbling in crossbreds using the optimum Wagyu sires in the outcross for optimum feeder performance.

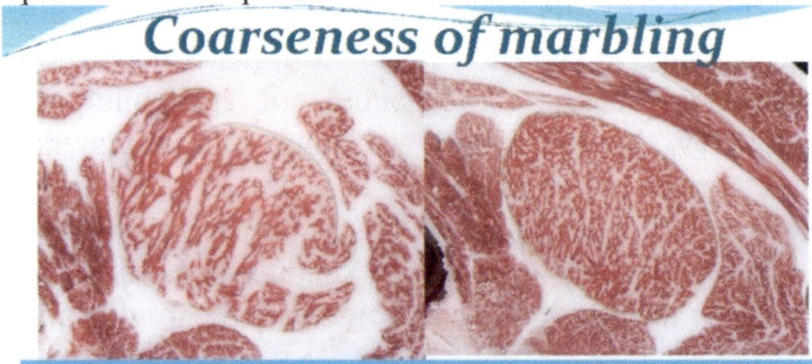

Japanese BMS number	6	7
Marblingfat%	58.6	57.7
Marbling coarseness %	35.9	4.4
Wholesale price yen/kg	1,850(e)	2,220(e)

Of particular interest was that part of the discussion led by **Peter Gilmour** on use of the digital camera to measure fine marbling - and how marbling (BMS) scores vary from the

Australian scores in Japan and USA. The importance of ensuring reproducible and accurate evaluation of the high value fine marbling score was repeatedly raised by discussants , suggesting that it has not always been the case (Digital camera shots at the abattoir should be taken at ≤ 4°C given the low Tm value of the intramuscular fat). Coarse marbling is to be avoided , particularly in the export market in Japan.

Keith and Alex Hammond presented two very interesting and definitive updates on their goals at Robbins Island Wagyu to produce polled full blood Wagyu and a high quality Wagyu x Akaushi cross. Considerable progress has been made. They both made a point of acknowledging their collaboration with CYO Foundation in the realization of their goals, particularly their practical awareness of the marbling features associated with the C19 60.1 MRIP haplotype.

Current Polled Animals - All Heterozygous Polled

- Flush cows, embryos
- Polled cows in calf and currently calving
- Bar R 52Y (Polled American Purebred Bull, 60.1)
- 18 polled yearling heifers from different line
- 4 scurred yearling bulls (one 60.1, 60.1)
- 3 year old scurred bull (carrying 60.1)
- This years calf crop (should get ~800 polled animals

Bar R 52 Y

Polled Wagyu: **Keith** took us through the main Robbins Island goals "To breed a herd of high percentage, homozygous polled, purebred Wagyu " In this way their aim is to produce for the commercial market both homozygous and heterozygous polled Wagyu bulls and thus lead the Australian Wagyu industry "towards polled cattle, whist maintaining high marbling and soft fat characteristics in the end product ". Indeed the results presented showed that Robbins Island was well advanced in producing polled Black Wagyu in the majority of their herd. In particular the polled phenotype is coupled in their bulls with the marbling 60.1 MRIP haplotype.

Wagyu (Black) x Akaushi (Red) Crosses: **Alex** outlined the plans and the advantages of infusing marbling and soft fat characteristics of Black Wagyu into the bigger framed and apparently aster growing and hardier Red Wagyu to produce a cross " .. that will have the Black Wagyu soft fat and marbling characteristics, combined with the growth of the Reds … (in the hope)… they will fatten off either grass or grain at a younger age and higher carcass weight than is typical of Black Wagyu cattle ." These are noble goals and is the approach that will be attempted also in the northern Kimberley. Indeed on Robbins Island, it is already clear that the Reds fattened on grass for their own use all have produced respectable BMS marbling scores ≥ 5 with high eating quality. This has been achieved with higher growth rates and better maternal characteristics than the black Wagyu. Particular attention was drawn to the superior daughters of Red sire Ashwood FX014 bred by **Bob Reed** at Mazda Wagyu in Victoria. (874, G878 - F X014 Daughters).

Alan Peggs representing **Pardoo Station** in the Kimberley purchased by **Bruce Cheung** earlier this year – a major collaborative CYO partner - outlined their Wagyu infusion plans. The goal, much like Robbins Island in the south to improve the eating quality of the Pardoo Santa Gertrudis herd, particularly for the Chinese retail market. Again Bob Reed's superior Akaushi semen is being employed.

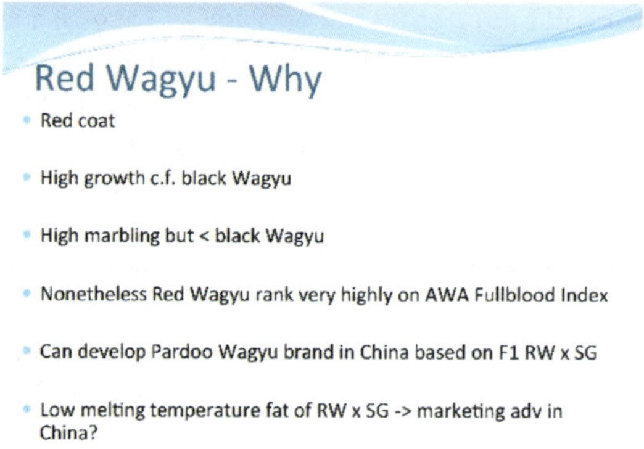

David Stoate of Anna Plains Station near Roebuck south of Broome outlined his reasons for trying to improve his Brahamn herd with Wagyu infusions (from Irongate Wagyu bulls). Their primary market is live export to the Middle East and to Indonesia. This market was severely damaged by the recent live export ban under the Gillard Government triggered by animal rights activists so are looking at diversification options. They run about 16,000 Brahman cattle over 380,000 hectares. David's family run business with wife Helen Campion will try Wagyu x Brahman on a small scale as they feel their way in these types of ventures. Read their interesting story at http://www.target100.com.au/Farmer-stories/David-Stoate-Helen-Campion

Bob Reed of Mazda Wagyu contributed throughout the discussions on the favoured 'marbling on grass features' of Akaushi. Their genetics are contributing to Akaushi herds and cross breeding on Melaleuka Stud, Robbins Island Wagyu and Pardoo Station.
Andrew Rado gave a lunch presentation of the plans of Stone Axe Pastoral . Together with **Mathew Walker** are setting up 6000 head full blood Wagyu operation in the heat belt at Kojunup all free of genetic test disorders. Mathew is the son of Chris Walker who started the original Australian commercial production of fullblood Wagyu into Australia (with the importation of 105 fullblood cows and bulls) known as Westholme Wagyu and sold to AACo.

The other main features of the meeting was a well attended **Charity Dinner** at the CYO Pub and Tavern in the evening which was attended in good spirit. On the Friday after a wrap up steered by Professor Dawkins on future collaboration plans the interstate visitors were drivenin the CYO Foundation minibus over the various **Melaleuka Stud** paddocks and properties at **North Dandalup (Mundup)**. In many ways

Melaleuka stud is one large experimental breeding farm where pure bred Simmental, Wagyu, Akaushi herds and crosses with Gelbvieh, Dexter and Black Angus with both grass paddocks and feed lots – the aim being to achieve the goals outlined by Professor Dawkins in the introduction – reductions of the days on feed ($<$ 150 days) to produce good weight gain and beef with low Tm intramuscular fat ($<$ 37°C) as illustrated in Figure 1.

Visitors

During the year (4 May 2015) a presentation on "Immunotherapy in Cancer" was presented by Professor Brendon J Coventry, Director, Immunotherapy Laboratory, Research Director, Australian Melanoma Research Foundation, Royal Adelaide Hospital, University of Adelaide.

Other visitors in March brought in by Mr Peter Gilmour, IrongateWagyu, Albany, included AWA Executive Officer Mr Graham Truscott and Wagyu Technical Officer and ABRI CEO Mr Alex McDonald.

On 13-15th August Sue Lester, from the Rheumatology Department Queen Elizabeth Hospital, Woodville, South Australia, set up training workshops in the R statistical package, to facilitate the analysis of genomic, Tm and carcass data in both Wagyu and non-Wagyu breeding and feeding programs.

Visit of Charles Sturt University Agribusiness Students to CYO Pub & Tavern

On Monday February 29 2016 a class of 25 first year students led by Adjunct Lecturer Wendy Dymond from the School of Agribusiness, a joint offering by C.Y. O'Connor Institute, Northam and Charles Sturt University (NSW) visited the CYO'Connor Pub and Tavern at Piara Waters between 1:00 - 3:00 PM. Most of the students were WA born and bred, and most were from farming families.
It was initially organized by Pub Manager Jason Hampton and Wendy Dymond around a series of brief presentations about the livestock research of the CYO Foundation and the

Paddock-to-Plate program with question and answer sessions followed by a blind steak tasting.
Mr Anthony Lloyd, CYO Foundation Vice-Chairman welcomed the students with homilies and essential lessons in farm 'economics.' In our assessment the students seemed very engaged (this was their first day at university).
Mr John Dawkins, Director of Melaleuka Stud and CYO'Connor Village Pub's Paddock to Plate Program gave a presentation which really engaged the student's interests – a straight talking outline of the economics and quality outcomes of a functioning commercial Paddock to Plate program. His talk drew probing questions , particularly on the Wagyu F1 program.

Dr.Ted Steele of CYO Foundation presented a brief talk on the science of the CYO Healthy Beef Project covering CYO's invention of the Rotorgene fat melting point (Tm) procedure and its application to beef fat Tm determinations in grain fed Wagyu v European breeds. Again a number of students engaged in scientific question and answers, and particularly about Wagyu outcrossing to improve the beef quality of *Bos indicus* tropical breeds. This led to discussions on the CYO collaboration with Pardoo Station in the Kimberly.

Ms Emily Tomsett the Village Green Beef Ambassador gave an informative talk about the CYO Pub & Tavern public education program with respect to creating greater awareness of Paddock to Plate quality beef cuts - and particularly the Wagyu and Wagyu F1 cuts of meet offered by the Tavern.

This was followed by a blind steak tasting with the students offering their scores on the tenderness, palatability and estimated Tm of the served steak.

Wendy Dymond thanked us on behalf of the students. In the CYO Foundation's assessment this was a very useful exercise with respect to future collaborations with CYO'Connor Institute-Charles Sturt University. This is particularly so as this target group of students could form part of future graduate students undertaking both work experience and livestock - agriculture research projects at CYO-Melaleuka Stud.

Optimism About Future of Degree at Muresk

To get a taste of what it is like to go to a regional university, attend the agribusiness taster at Muresk on either 21 and 22 April or 28 and 29 April. Details on 9622 6789

Printed in Dunstable, United Kingdom

91222399R00047